The Late Holocene Geomorphic History of Montezuma Canyon, Southeastern Utah, and the Puebloan Agricultural Landscape

Wayne K. Howell and Eric R. Force

Arizona State Museum
THE UNIVERSITY OF ARIZONA.

Arizona State Museum Archaeological Series 213

Arizona State Museum
The University of Arizona
Tucson, Arizona 85721-0026
Copyright © 2018 by the Arizona Board of Regents
All rights reserved.
Printed in the United States of America

ISBN (paper): 978-1-889747-99-6
Library of Congress Control Number: 2017963814

ARIZONA STATE MUSEUM ARCHAEOLOGICAL SERIES

General Editor: Richard C. Lange
Technical Editor: Kelly M. Alushaj

The *Archaeological Series* of the Arizona State Museum, The University of Arizona, publishes the results of research in archaeology and related disciplines conducted in the Greater Southwest. Original, monograph-length manuscripts are considered for publication, provided they deal with appropriate subject matter. Information regarding procedures or manuscript submission and review is given under Research Publications on the Arizona State Museum website: *www.statemuseum.arizona.edu/research/pubs*. Information may be also obtained from the General Editor, *Archaeological Series*, Arizona State Museum, P.O. Box 210026, The University of Arizona, Tucson, Arizona, 85721-0026; Email: langer@email.arizona.edu. Electronic publications and previous volumes in the Arizona State Museum Library or available from the University of Arizona Press are listed on the website noted above. Print-on-demand versions of the lastest Arizona State Museum Archaeological Series may be obtained from several booksellers on-line.

The Arizona State Museum *Archaeological Series* is grateful to the many donors and supporters who continue to make this publication possible.

FRONT COVER: Remains of a Pueblo III unit pueblo exposed on surface of alluvial fan at Horsehead Canyon. Photograph by Eric Force.

Contents

Contents	iii
List of Figures	v
List of Tables	vi

Chapter 1: The Study	1
Setting	1
Regional Research Context	2
Previous Research in Montezuma Canyon	9
Methods	10
Chapter 2: Depositional Units, Stratigraphy, and Geomorphology	13
Pearson Unit	14
General Description	14
Facies	15
Age	16
Fluvial Paleo-environment	17
Horsehead Unit	17
General Description	17
Facies	18
Age	19
Fluvial Paleo-environment	19
Nancy Patterson Unit	20
General Description	20
Facies	20
Age	20
Fluvial Paleo-environment	21
Coalbed Unit	22
Stratigraphy and Geomorphology	22
Chapter 3: Archaeological Relationships	25
Panorama Point Locality	25
Montezuma Village, Upper (MC5) and Lower (MC15) Localities	27
Horsehead Canyon Locality (MC13)	29
Coalbed Village Locality (MC8/MC11)	31
Rattlesnake Site Locality (MC10)	34
Bradforn Canyon Alluvial Fan Locality (MC9)	36
Three Kiva Pueblo Locality (MC6)	37
Tank Canyon Locality (MC4)	41
Cave Canyon Locality (MC16/MC17)	43
Bighorn Site Locality (MC3)	43
Nancy Patterson Village Locality (MC1/MC2)	45
Summary of Depositional and Archaeological Relations	47
Chapter 4: The Puebloan Agricultural Landscape	49
The Early Puebloan Landscape	50
The Late Puebloan Landscape	50

Chapter 5: Conclusions and Implications ... 53
    Findings on Geomorphic History and Puebloan Agriculture ... 53
    Lowland Puebloan Historical Ecology and Population Dynamics ... 54
    Implications for Management and Future Research ... 56

References Cited ... 59

Figures

1.1	Montezuma Canyon in regional context	2
1.2	Montezuma Canyon and its main tributaries, with study sites labeled "MC" and numerically listed in sequence of discovery	4
2.1	Christensen's type for his Unit 1 and Unit 2, and our Pearson type site	13
2.2	Pearson Unit exposure at MC15	16
2.3	Horsehead Unit type site	18
2.4	Nancy Patterson Unit type site	21
2.5	Schematic cross-section showing stratigraphic relationships of the four alluvial units	23
3.1	Aerial view of MC15 study site below Pearson Canyon	26
3.2	Isolated plug of the Pearson Unit at MC15	27
3.3	Map of Montezuma Village showing location of the MC14 study site	28
3.4	Aerial view of the confluence of Hosrsehead Canyon and Montezuma Canyon at MC13	30
3.5	Two sandstone blocks in profile	31
3.6	Aerial view of Coalbed Canyon vicinity	32
3.7	Rubble mounds of lower Coalbed Village	34
3.8	Aerial view of the Rattlesnake Site vicinity	35
3.9	Profile showing the relationship of the Pearson and Horsehead units at the Rattlesnake Site	36
3.10	Facies profile at the road crossing of the Bradford Canyon stream	37
3.11	A Pueblo III-age roomblock exposed in the east cut bank of the county road at Bradford Canyon	38
3.12	Aerial view of the Three Kiva Pueblo vicinity	39
3.13	Profile showing the relationship of the Pearson and Horsehead units in the middle canyon floor	39
3.14	A Pueblo I roomblock is visible as rubble and several rows of upright slabs	40
3.15	A Pueblo II unit pueblo has been bisected by modern erosion	42
3.16	An arroyo has bisected a compound terrace in the Bighorn Site area	45
3.17	Pieces of charred wood are common in inter-bedded clay and sand channel fills	46
4.1	A possible cultural landscape during Basketmaker III and Pueblo I times	51
4.2	A possible cultural landscape during Pueblo II-Pueblo III times	52

Tables

1.1.	Pecos Classification System	8
2.1.	Radiocarbon Dates	14

Preface and Acknowledgments

The seed of this study was planted during the summer of 1977 when Wayne Howell was a young archaeology student at Brigham Young University's field school working at Cave Canyon Village in Montezuma Canyon. While picking through the architectural details of a Pueblo I pithouse/proto-kiva that dated to the early 900s, it was difficult to imagine how anybody could have made a living farming the eroded, desiccated bottomlands nearby. Later, after excavating at upland Pueblo I villages near the Dolores River valley and on Mancos Mesa in southwestern Colorado, it became clear that the answer as to understanding how the Ancestral Puebloans could have lived in such altogether different habitats lay in understanding the floodplain history of the lowlands where agriculture would have been such a challenge. The key to deciphering that mystery for Howell happened in 1992 when Eric and Jane Force decided to pay a visit to Kelly Place in McElmo Canyon just outside Cortez, Colorado. Force, a retired USGS geologist, was intrigued by a profile exposed in a partially excavated kiva at Kelly Place, which showed it had been buried by flood deposits. He asked around and was put in touch with Dr. William Lipe at Crow Canyon Archaeological Center in Cortez. Lipe saw the potential of a collaborative research project and brought Force and Howell together. The result of that study was *Holocene Depositional History and Anasazi Occupation in McElmo Canyon, Southwestern Colorado*, published in the Arizona State Museum Archaeological Series 188. This study is an outgrowth of that effort.

Although this project was undertaken independently by the coauthors, it occurred under the umbrella of the decades-long research carried out in Montezuma Canyon by the Brigham Young University field schools directed by Dr. Ray Matheny and later by Dr. Joel Janetski. This study was initiated because Dr. Matheny and his wife, fellow archaeologist and Montezuma Canyon alumnus, Dr. Deanne Gurr Matheny, undertook a project to synthesize the decades of research they and other BYU archaeologists have carried out in Montezuma Canyon since the 1960s. Without the Matheny's inspiration and dogged determination to complete this lifetime of work, this study would have never occurred.

Field work was initiated in 2012 by the authors with a field inspection in the company of Ray and Deanne Matheny, Dr. Fumiyasa Arakawa with New Mexico State University, Dr. James Allison with Brigham Young University, Don Simonis with the Bureau of Land Management, Monticello District, and Marcia Simonis and Jane Force. With assistance from a grant provided by the Charles Redd Center for Western Studies at BYU, the Forces and Howell were able to return to the canyon in 2013 to expand the number of study sites and sort out the complex array of depositional packages and archaeological site associations they had observed the year before. As analysis continued and new questions arose over interpretations, Howell was able to return to the canyon in subsequent years to clarify unit associations and archaeological site ages. He was accompanied on these field excursions by archaeologist Winston Hurst, of Blanding, Utah, who provided expert ceramic analysis, and variously by artists Joseph Packak and Kyle Bauman of Bluff, Utah. Photographs used in this publication were provided by Kyle Bauman (Fig. 3.16), Eric Force (Fig. 2.2, 2.3, 3.2, 3.5), Wayne Howell (Fig. 3.10), Winston Hurst (Fig. 3.11), Deanne Matheny (Fig. 2.4, 3.7, 3.17) and Joseph Pachak (Fig. 2.1, 3.14, 3.15). The map of Montezuma Village (Fig. 3.3) was created by Sarah Baer, as adapted from the original by Ray Matheny. Sketches were done by Eric Force (Fig. 3.13) and Wayne Howell (Fig. 2.5, 3.9, 4.1 and 4.2). Thanks to Kennth Wintch and Barry Biediger with Utah Trust Lands Administration for assisting in getting the aerial imagery together, all of which was downloaded from Google Earth. Figure 1.2 was provided by the Utah Automated Geographic Reference Center though UTLA. Francis Broderick of Arch Graphics in Anchorage, Alaska, is responsible for design and production of all of the figures. A thanks to the Bull family, in particular Danny Bull, for allowing us access to their property in upper Montezuma Village, and to Howard and Lyla Ransdell for providing access to their property in the heart of Montezuma Village. A special thanks to Liza Doran at Cow Canyon Trading Post in Bluff, Utah, for hosting participants during the first expedition, and a heartfelt thanks to Kim Ney for her patience and endurance as this study has lurched to completion.

Chapter 1
The Study

The purpose of this study is to examine the timing and ways in which fluvial processes shaped the Montezuma Canyon alluvial plain into the landscape that provided the agricultural foundation for Ancestral Puebloan people. This includes from the time they migrated into the area, about the beginning of the common era, until their abandonment some thirteen centuries later, and focuses on how the alluvial plain evolved through time and how the Puebloans adapted their settlement system to this changing lowland environment. This understanding is particularly relevant because Montezuma Canyon is recognized as having supported substantial Puebloan populations in the past, yet the floodplain of Montezuma Canyon lies below the elevational gradient where dryland farming in the northern San Juan region — considered the mainstay of Mesa Verdean Puebloan subsistence — is considered feasible (Bocinsky and Kohler 2014; Petersen 1988; Van West 1994; Varien 2002; Varien et al. 2007). This means the Montezuma Canyon Puebloan agricultural system would have been entirely reliant on runoff channeled into the canyon from surrounding uplands. Where and how this runoff flowed and shaped the sediment packages determined the Puebloan agricultural landscape, just as elevation-driven rainfall patterns and soil distributions determined it in the uplands.

Montezuma Canyon is an ideal setting for this type of study because of its rich archaeological heritage and the fact that the relationship of its archaeological sites to the deep sedimentary deposits that fill much of the canyon was exposed by erosion sometime around A.D. 1900, revealing a shared stratigraphic history. As currently exposed, the canyon bottom's geomorphic history of interest begins in the middle Holocene period, sometime around 6000 years ago, and contains much fine-grained detail during the Puebloan period of occupation, from about A.D. 1 to 1300, and extends to the modern era.

This multi-disciplinary approach is synergistic in that it uses geological techniques to understand the fluvial processes that shaped the floodplain through time and it interprets the stratigraphic placement of Puebloan cultural remains in relation to sedimentary units as a tool for constructing fine-grained chronological controls for determining the relative ages of the depositional packages. Concomitantly, these sedimentary relationships help us to understand when and how people adapted their settlement and agricultural practices to the floodplain.

Setting

Montezuma Canyon is the central drainage of the Blanding Basin, a broad structural basin draining the north central portion of the San Juan River drainage (Figure 1.1). The San Juan River arises in the San Juan Mountains

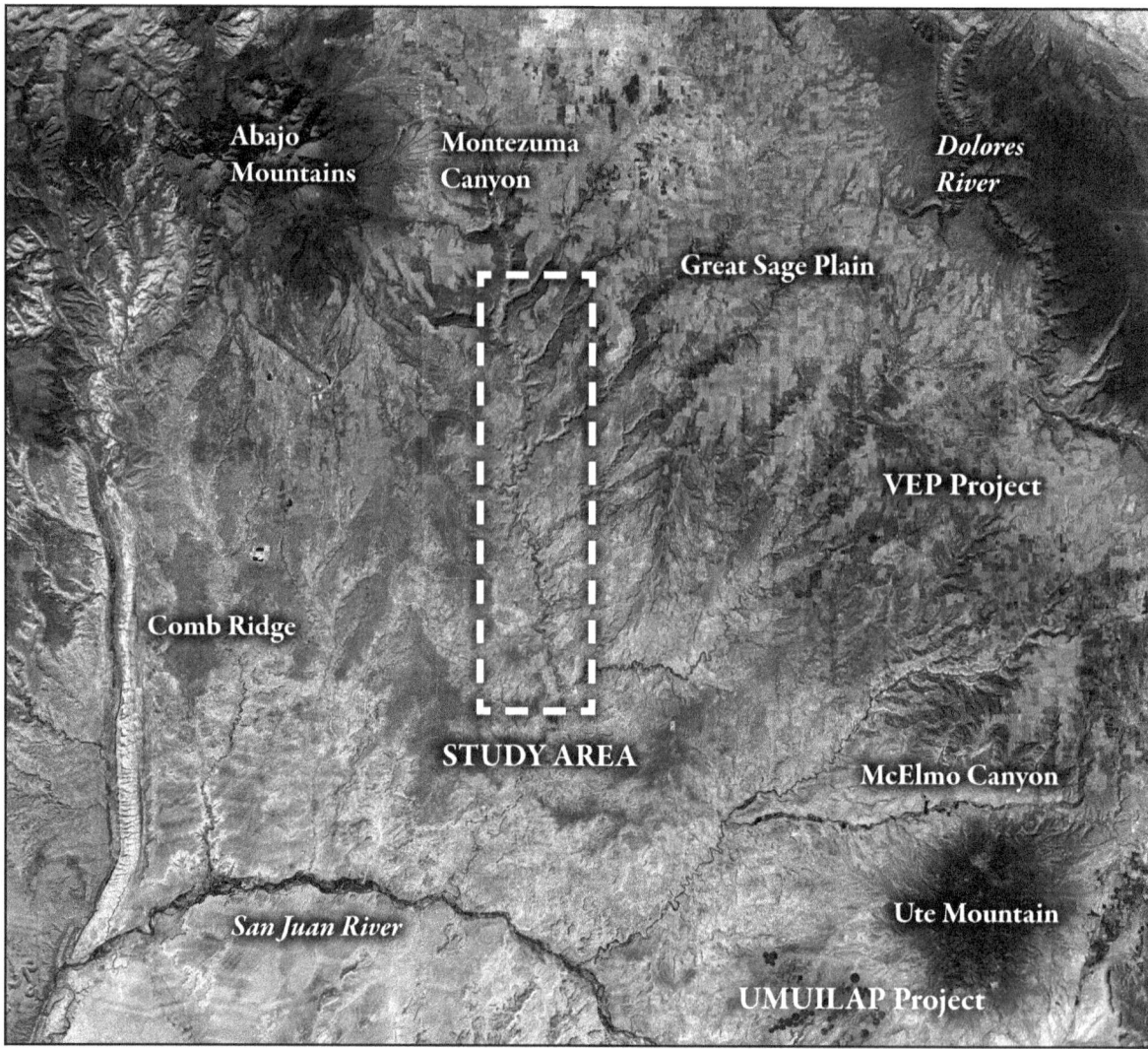

Figure 1.1 - Montezuma Canyon in regional context. Montezuma Canyon is the central drainage in the Blanding Basin, the fan-shaped basin north of the San Juan River. Two archaeological studies referenced in the text — VEP and UMUILAP — are also indicated. Image taken from Google Earth, 2017.

in Colorado and cuts from southeast-northwest through the Blanding Basin on its way to the Colorado River. The Blanding Basin forms a fan-shaped arc of five drainages that channel runoff generally from north-to-south toward the San Juan River. Comb Ridge defines the basin on the west and the McElmo Dome and Great Sage Plain Plateau on the east and north. The igneous-intrusive mass of the Abajo Mountains feeds the headwaters of four of the basin's drainages, including Montezuma Canyon which is the largest and easternmost of the four.

The Montezuma Canyon drainage basin encompasses some 3106 km (1200 mi) with a main channel that extends over 122 winding km (76 mi) from its headwaters in the Abajo Mountains to its confluence with the San Juan River. The Montezuma Canyon drainage basin also has great breadth, with its west side tributaries reaching into the Abajo Mountains and its east side tributaries extending through the Great Sage Plain and virtually to the rim of the Dolores River canyon in southwestern Colorado, a linear distance of over 66 km (42 mi). The drainage also has a considerable vertical reach, with its main stem headwaters falling from an elevation of over 3445 m (11,360 ft) in the Abajo Mountains and dropping to 1341 m (4400 ft) at Montezuma Creek on the San Juan River, a range of almost 2100 m (7000 ft). Many of the canyon's tributary drainages flow from upland plateaus that rise on average between 1829 and 2134 m (6000 to 7000 ft). Montezuma Canyon is incised into the surrounding countryside on average about 300 m (1000 ft), so for most of its course the canyon bottom lies below 1700 m elevation (5500 ft). Within much of our study area the canyon has a seemingly flat valley floor, generally 0.5 to 1.0 km wide and meandering around bedrock promontories. The modern creek in our study area (Pearson Canyon to Cross Canyon) descends from about 1675 to 1450 m (5500 to 4750 ft) in elevation, a gradient of about 4.5m per km (24 feet per mile).

This great elevation range means that Montezuma Canyon also experiences a range of climatic conditions, and the topographic relief of the canyon determines where different rainfall patterns prevail. Bedrock geology is the determining factor for this topography, creating descending elevational gradients from north to south along the canyon's length (for more detailed geologic descriptions see Huff and Lesure 1965 and Christenson 1985). The highest elevation is composed primarily of Tertiary-aged igneous rocks of the Abajo Mountains that give rise to the canyon's main channel headwaters. This igneous laccolith was intruded into the surrounding Mesozoic rocks some 27 to 28 million years ago (Armstrong 1969), and being the most erosion resistant rock type in the drainage basin creates the greatest topographic relief. The rocks underlying the Great Sage Plain, which arcs around the head of Montezuma Canyon and abuts the base of the Abajo Mountains, are comprised of Cretaceous-aged sandstones, conglomerates, mudstones, and limestones of the Burro Canyon Formation and the Dakota Sandstone. These rock types are resistant to erosion and, where exposed, form the outer rim of the Great Sage Plain Plateau, which also effectively forms the upper rims of Montezuma Canyon and its tributaries. As these rims have eroded, the Upper Jurassic Morrison Formation has been exposed and it is the principal unit that forms most of Montezuma Canyon's walls and inner benches. This four member unit is comprised of inter-bedded layers of sandstones, mudstones, and conglomerates that vary in their resistance to erosion. These units, in addition to the silty loams that mantle the upland mesas, provide the source materials for the alluvial sediment packages we study.

This erosional variability has created two very distinct topographies in Montezuma Canyon. About the mid-point of our study area where Monument Canyon flows in from the east, the character of the canyon changes (Figure 1.1, see also Figure 1.2). Above that point, the bedded sandstones of the Morrison formation overlay the older Summerville Formation and, in the upper canyon, the underlying Entrada Sandstone is exposed to form a deep, meandering inner canyon with steep sides rising to benches, canyon rims, and the nearby upland mesa tops. We refer to this reach as the upper canyon, which, in most places, is 305 to 335 m deep (1000 to 1100 ft), with the linear distance from rim to canyon bottom in the range of 1600 m (1 mi) or less. Below the Monument Canyon junction, the less

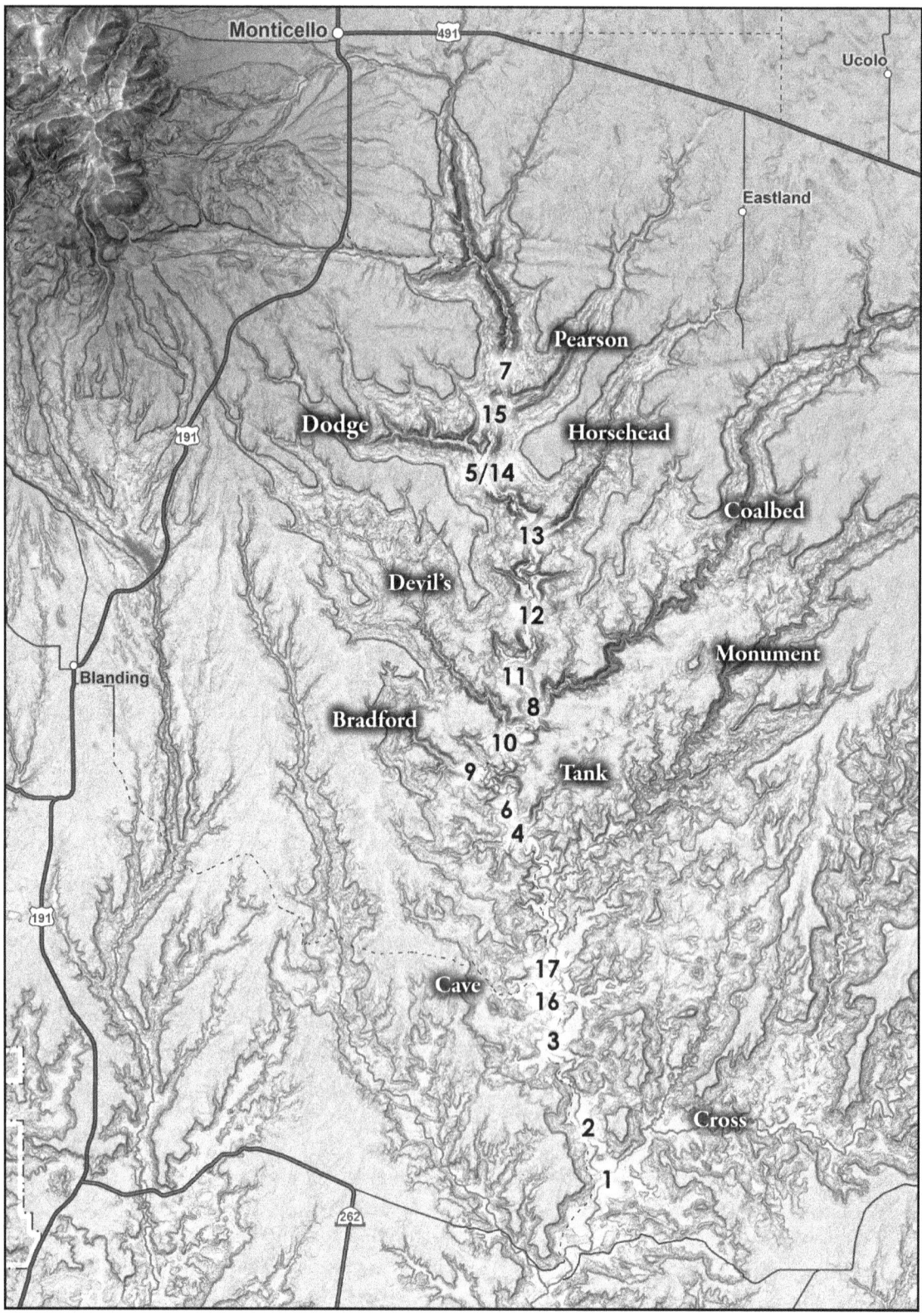

Figure 1.2 - Montezuma Canyon and its main tributaries. Study locations, labeled "MC" in the body of the report, for Montezuma Canyon, are numbered in sequence of discovery. What we consider the "upper canyon" extends from MC4 northward; the "middle canyon" extends from MC4 southward to MC1.

resistant rocks have eroded back from the main drainage and the canyon is wide and shallow and dominated by badland-type topography. Canyon rims are distant and upland mesas supporting pinyon-juniper forest are often many kilometers away. We refer to this as the middle canyon. The lower canyon stretches from Cross Canyon to the confluence with the San Juan River, but is outside our study area and is not discussed further in this report.

This great variability in topographic relief affects rainfall in a significant way. At the canyon's headwaters in the Abajo Mountains, rainfall can exceed 76 cm (30 in) of precipitation per year, including significant snowpack. On the upland mesas around Monticello and the Great Sage Plain, precipitation averages about 40.6 cm (16 in) per year. However, as one moves progressively south and away from the uplands, precipitation drops; at Hovenweep, located at about 1580 m (5200 ft) elevation on Cajon Mesa to the east of the middle canyon, average precipitation is about 29 cm (11.4 in) per year, and at Montezuma Creek on the San Juan River, it is down to about 20 cm (8 in) per year. With precipitation falling mainly in the form of winter snow and late summer thunderstorms, the streams in the upper canyon traditionally flowed for much of the year (less so since upland water users now divert those flows), whereas, those in the lower canyon traditionally flow only during periods of snow melt in spring or at times of heavy rainfall.

We found that the geographic subdivision of upper and middle canyon also proves useful when discussing the Holocene deposits that are most pertinent to our study. In the upper canyon, with its relatively narrow bottom and steep bedrock walls, sediments have been contained, valley fill is deep, and tributary canyon alluvial fans are pronounced. The modern stream has entrenched 10 m or more into this valley floor exposing the Holocene deposits, generally in nearly vertical banks. The relationships of sediment facies are well exposed and open to study. In contrast, in the middle canyon, the bottom is wide and sediments have been dispersed over a broad area. The modern stream has cut exposures that are generally not more than a few meters deep and sediment facies are not as exposed or discernible as they are up-canyon, making interpretation more challenging.

Soils and their suitability for agriculture formed the foundation of the Puebloan agricultural landscape. The Montezuma Canyon floodplain has five soil complexes that form alluvial fans, floodplains, and stream terraces, and are comprised of fine-to-coarse sandy loams and silt loams derived from the deterioration of sandstones and shale parent materials. Some soil types are classified as "farmland of statewide importance" provided they can be irrigated, but most are classified as "not prime farmland" (*http://websoilsurvey.nrcs.usda.gov*). Sand dunes, which helped support Puebloan agricultural systems in lowland areas such as the Hopi mesas and along the San Juan River benches in nearby Bluff, are absent to uncommon in the study area and would not have contributed significantly to the Puebloan agricultural system and are not further discussed.

Rainfall estimates for the canyon bottom range from 20.3 to 30.5 cm (8 to 12 in) in the upper end of our study area grading down to 15.2 to 22.9 cm (6 to 9 in) in the lower end. Although there are sufficient frost-free days throughout the project area to sustain agriculture — from 140 to 180 days — dryland agriculture would have been only marginally feasible even during the best of times in the uppermost reaches of our study area (Montezuma Village and above), but untenable down valley from that. Hence, the Puebloan populations who built and occupied the many farmsteads and villages along Montezuma Canyon had to rely entirely on floodwater farming as the only viable alternative.

REGIONAL RESEARCH CONTEXT

The Puebloan people of the American Southwest have a long history of migration, as expressed in their oral traditions and as revealed profusely in the archaeological record. Rich data provided by tree rings and the pollen record have provided the Southwest with a fine-grained, chronologically controlled climate record which has facilitated increasingly refined research into climate change as a primary causative factor for explaining local and regional population shifts, particularly in the northern San Juan region (Benson and Berry 2009; Berry 1982; Betancourt and Van Devender 1981; Bocinsky and Kohler 2014; Cordel 1975; Dean 1988; Dean and Van West 2002; Kohler 1992; Kohler and Varien 2012; Lipe 1992; Schlanger 1988; Schwindt et al. 2016; Van West 1994; Van West and Lipe 1995; Varien et al. 2007).

The linkage between the effects of rainfall and temperature on Puebloan agricultural potential — specifically as it affects the cultivation of maize — has focused research on changing precipitation and temperature patterns as factors forcing populations shifts, with regional droughts being perceived as the main drivers. Yet, for the most part research has tended to look at Puebloan agricultural responses to changing conditions from an either/or perspective; Puebloan farmers were either rain-fed dryland agriculturalists responding to changing conditions in the elevationally-constrained upland zone (between about 1700 and 2400 m), moving up or down gradient depending on changing rainfall or temperature patterns within a specific geographic area, or they were floodwater agriculturalists living in lowland settings and responding to climate induced incision of floodlands which affected agricultural potential (Bryan 1941; Dean 1988; Dean et al. 1985; Karlstrom and Karlstrom 1987). Although there have been only a few attempts to examine the Puebloan agricultural practice as an integrated bi-modal upland/lowland strategy (Dean et al. 1985; Dean and Van West 2002; Euler 1988; Euler et al. 1979; Hack 1942), most research efforts in the northern San Juan region have focused almost entirely on the uplands.

For example, over the past decade, researchers affiliated with the Crow Canyon Archaeological Center and Washington State University, using a strategy developed for the Village Ecodynamics Project (VEP I), have refined and integrated data on climate, soils, site settlement histories, population estimates and behavioral responses, and developed computer generated ecological models that allow for testing of assumptions on Puebloan responses to the changing ecology (Bocinsky and Kohler 2014; Kohler et. al 2007; Varien et al. 2007; Kohler and Varien 2012). This research has focused on a 1600 km² area in Montezuma County, Colorado (just east of our study area) and analyzes data on climatic factors on upland valleys and mesas extending from below the 2,400 m elevation contour (7900 ft) — the upper threshold of low growth degree days which precluded maize cultivation — to an area just east of the Colorado-Utah state line.

The research area and site database was expanded by the VEP II study to model what the authors termed the "maize niche" (Schwindt et al. 2016) refining the conclusions of the VEP

I study. This upland area supported rainfall-dependent dryland farming during climatically favorable periods. Climatic variation that precluded dryland farming in those uplands is posited as a cause for out-migration to more favorable environs on the Colorado Plateau such as Chaco Canyon or the Pajarito Plateau (Bocinsky and Kohler 2014), or off the Colorado Plateau and into the Sonoran Desert (Benson and Berry 2009; Varien et al. 2007). However, these modeling exercises end at the 30 cm of annual precipitation isohyet contour below which rainfall dependent maize cultivation would have been precluded — about the elevation of Hovenweep (1580 m and 29 cm precipitation) — but do not account for the ability of the Puebloan farmers in areas such as Montezuma Canyon to persist in place for generations in a lowland setting.

The inability to model the agricultural potential of the Puebloan lowlands is not surprising given our fragmentary and inconsistent understanding of the dynamics that drive those systems, although recent research is beginning to address this shortcoming (Eiselt et al. 2017). The long-standing notion that regional droughts prompted a chain-reaction — drought depressed the water table which caused arroyo cutting which led to loss of agricultural potential which in turn drove local and regional abandonments (Antevs 1952; Bryan 1929, 1941; Euler et al. 1979; Haynes 1968; Karlstrom 1988) — has largely been laid to rest (see Huckleberry and Bilman 1998 for an excellent synopsis of the contentious "arroyo climate debate"), although elements of it persist in current dialogue (Benson and Barry 2009). Although arroyo-climate studies provided many valuable insights, and a general correlation in the timing of certain drainage-to-drainage erosional events is defensible, consensus on a region-wide chronostratigraphic sequence never congealed. This is perhaps because the direct link between drought and down-cutting remained tenuous and there was no consensus on the geomorphic processes involved. This owes largely to an evolving understanding of what those processes were. For example, Karlstrom (1988) considered the climate-induced fluctuation in the water-table as a variable independent of local conditions such as drainage physiography and sediment load. However, Cooke and Reeves (1976) observed that, in any given drainage and at any given time, the locus of entrenchment occupies only a portion of a valley segment, with sediments dislodged by incision in-turn deposited down-drainage. Schumm (1977) and Patton and Schumm (1981) noted that discharge rates may fluctuate from decade to decade, but sediment load can remain constant as the loci of sediment storage migrates headward in a system. They termed this process "complex response," with individual drainages evolving independently based on local variables.

Our understanding of how the process played out in the northern San Juan drainage has been informed by two studies carried out near Montezuma Canyon during the mid-1990s. Eric Force and Wayne Howell examined a 6.5 km (4 mi) stretch of middle McElmo Canyon where excellent exposures revealed numerous Puebloan archaeological sites in various contexts arrayed upon and within Holocene alluvial deposits (Force and Howell 1996). Working on a landform they called the "Anasazi terrace," they identified two main depositional units which formed the habitation landscape throughout Puebloan time. The older unit is comprised of alluvial sediments that filled the main valley and represents a meandering channel-floodplain deposit that was greatly influenced by alluvial fans discharged from side drainages, most prominently from north-side canyons channeling flow off the McElmo Dome. This aggrading alluvial plain was inhabited by Basketmaker III (BMIII) people (Table 1.1), as their cultural remains were found to be resting upon the valley surface or in a few cases embedded in northside fans. This long-stable landform was cut by an entrenchment event that appears to have occurred primarily in late Pueblo I (PI) time, which created badland topography with steep slopes and vertical cutbanks in many places. However, aggradation followed shortly on the heels of incision and, during Pueblo II (PII) time, the inner valley arroyo and the newly aggrading north-side fans became an attractive environment for Puebloan agriculturalists who established residences on stable landforms along the valley margins. As the aggradation node moved up-valley, Puebloan settlement appears to have moved along with it.

A key finding of this study was that while entrenchment might have been of local importance in reducing or eliminating agricultural potential, its effects were short-lived and mitigated by the Puebloan agriculturalists' ability to adapt their settlement strategy to track the diachronous nature of the entrenchment/deposition progression. This study highlighted the need to improve our understanding of local floodland evolution as it relates to agricultural potential and of population mobility and migration in response to it. While Force and Howell agreed that drought may have been a factor contributing to the main incision event, they did not feel that entrenchment would have been a cause for Puebloan abandonment of the short reach of McElmo valley bottom they studied, much less the whole region (Force and Howell 1996:36).

Concurrently, Gary Huckleberry and Brian Billman were conducting similar work on the Cowboy Wash drainage in the Ute

Table 1.1. Ancestral Puebloan Phases Established by the Pecos Classification System, Revised for the Mesa Verde Region.

PECOS Phase	Time Period
Basketmaker II	500 B.C to A.D. 500
Basketmaker III	A.D. 500 to 750
Pueblo I	A.D. 750 to 900
Pueblo II	A.D. 900 to 1150
Pueblo III	A.D. 1150 to 1300

Mountain Irrigated Lands Archaeological Project (UMUILAP, see Figure 1.1) on the piedmont on the southern flank of Sleeping Ute Mountain (Huckleberry and Billman 1998). The UMUILAP project tested the two main schools of thought regarding stream entrenchment – 1) that most episodes of erosion result from climate change and its effects on vegetation cover and sediment yield (the arroyo-climate model), and 2) adjustments in a fluvial system may be in response to discharge, sediment load and channel geometry but not necessarily to regional climate change. While they found value in both schools of thought, they were unable to correlate their five periods of Puebloan occupation and two of abandonment in their study area with cycles of arroyo incision, and could not correlate the eight or nine periods of drought evident in the regional tree-ring series with arroyo behavior in their study area.

They acknowledge that the main episode of valley entrenchment may have occurred during PI times as noted by Fuller (1988) (and similar to what Force and Howell noted in McElmo Canyon), but they caution against attempting regional correlations. They further argue against "drought-induced stream entrenchment as a driving force in episodic regional abandonments" (Huckleberry and Billman 1998:597), and argue instead for study of each drainage's history independently, which in the case of Cowboy Wash, they define as a "discontinuous ephemeral stream" that occasionally provided conditions favorable for Puebloan ak-chin farming strategies (Huckleberry and Billman 1998:611), with active farming on aggrading channel fans and at side-channel tributaries, again similar to what Force and Howell had noted in McElmo Canyon.

They consider that the Ute Mountain piedmont was habitable during times of favorable regional climatic conditions (generally synchronous with the uplands where dryland farming prevailed) and may have been periodically abandoned because of the relatively small size of the watershed, making it vulnerable to extended periods of drought (Huckleberry and Billman 1998: 611). Huckleberry and Billman recognize that floodwater farming in several of the larger drainage systems in the Southwest could support substantial populations for extended periods, such as at Chaco Canyon and the Tucson Basin (Fish et. al 1994), a factor which argues for studies in larger drainages with long occupation histories. This study of Montezuma Canyon provides an example of a larger system for the northern San Juan basin, but one lacking a perennial river such as the Animas, La Plata and Dolores river valleys.

Previous Research in Montezuma Canyon

Two previous geologic studies carried out in Montezuma Canyon provide excellent background information for understanding this complex system. In a study of uranium deposits carried out in upper Montezuma Canyon in 1956 (Huff and Lesure 1965), Frank Lesure recognized two distinct depositional units filling the main valley; an older floodplain deposit which he was able to determine had filled the entire valley bottom and lapped against the valley walls, and a younger sandy deposit that overlay it in both main channel and fan facies contexts. Lesure was able to draw profiles of four sections, which he shared with us, and we were able to relocate and examine two of them. Lesure was able to demonstrate the utility of Puebloan architectural remains for relative dating of depositional units, as he found several masonry structures resting on their surfaces.

As part of a larger study of Quaternary deposits on the Colorado Plateau carried out by the Utah Geological and Mineral Survey in the 1980s, Gary Christenson conducted a detailed study of the geomorphic histories of Montezuma Canyon and nearby Recapture Canyon (Christenson 1985). Christenson was able to describe Montezuma Canyon's history from initial incision to the present, identifying three Quaternary-aged depositional units. Gravels of Pleistocene age were deposited on terraces etched onto the canyon's slopes at various elevations with continued incision carving the inner gorge by late Pleistocene time. Early Holocene deposits are lacking, but by the mid-Holocene, sometime prior to about 5500 B.P. (based on one radiocarbon date), Christenson was able to identify a floodplain deposit comprised of sands, silts and clays developing to a depth of more than 15m (50 ft) in places. This unit was incised by deep arroyos that were later in-filled by sands and gravels during an episode which Christenson interpreted as having occurred sometime prior to 1430 B.P. based on two radiocarbon dates collected from a profile north of the mouth of Pearson Canyon.

Christenson also recognized the relationship of geomorphic history to Puebloan settlement and provided a succinct summary of his interpretation of their relationship — that the main erosion event likely occurred prior to Puebloan settlement, and that the absence of Basketmaker-aged sites on the valley floor, where later Puebloan sites were present, may be indicative of the relative age of those surfaces (Christenson 1985:22-23). Our study builds upon Christenson's geological study and refines his geomorphic descriptions but provides a different interpretation of both the timing of the incision event as well as how the Puebloan settlement history related to it.

Prior archaeological observations in the study area began with the entry of the first exploring party to enter the San Juan Basin in 1875 under Ferdinand Hayden (Jackson 1878). The observations of photographer W.H. Jackson detail the abundance of cultural remains in the canyon as well as the presence of a perennial stream. Subsequent research by Mitchell Prudden, Edgar Hewitt, and Byron Cummings added to our understanding of the canyon's archaeology, but we rely most heavily on the work carried out by Brigham Young University under the direction of Ray Matheny during the 1960s to 1970s and by Joel Janetski in the 1980s. These studies provide survey data for sections of the upper and middle canyon and detailed settlement histories for four sites in our study area — Three Kiva Pueblo, Monument Village, Cave Canyon Village, and Nancy Patterson Village (Baer 2003; Billat 1985; Christensen 1980; de Haan 1972; Harmon 1977, 1979; Hurst and Janetski 1985; Janetski and Thompson 2012; Matheny 1962; B. Miller 1976; D. Miller 1974; Nielsen 1978; Patterson 1975; Wintch 1990).

METHODS

Though admittedly ambitious, this study has many limitations. It is foremost a rapid reconnaissance of a very large and complex system. Approximately 20 field days are represented here, clearly insufficient time to examine the entire reach of canyon from its headwaters to its junction with the San Juan River. Recognizing this, we chose to focus our attention in the area of canyon with the best exposures and the greatest concentration of archaeological sites, which is also where Brigham Young University archaeologists focused most of their research over 25 years. This area offers many good vertical exposures that are easily accessible from the road. We made observations along a stretch from the junction of Pearson Canyon with the main stem of Montezuma Canyon down to Cross Canyon, but our greatest density of observations extend from Pearson Canyon down to Tank Canyon. Even in this stretch — 35 straight-line kilometers — parcels of private property prevented any observations except from afar or by air-photo (Google Earth and Bing Maps proved invaluable in this regard). Below Tank Canyon we lost good exposures to examine and, by Monument Canyon, the nature of the canyon begins to change; it opens up considerably and the depositional facies (distinctive depositional units), so recognizable up-canyon, become less distinct and traceable. We found 17 sections where we were able to identify the relationships of depositional units to each other and/or the relationships of cultural sites to those deposits. Their locations are shown in Figure 1.2.

One other limitation we faced was our inability to establish absolute dates for sedimentary units. We rely on three radiocarbon dates collected by Gary Christenson (1985:20) for establishing the general ages for the two older Holocene depositional units. However, our concern over the "old wood" problem in radiocarbon dating tempers our reliance on interpreting Christenson's dates as absolute. For one, his technique of gathering contextually dispersed charcoal of unknown species from a general depositional context requires broad latitude in interpretation.

In an environment where trees such as Utah juniper and pinyon pine can reach average ages of 400 to 800 years, but may live beyond 1000 years or older, and where their charcoal can persist indefinitely in open-air or buried contexts, we caution against relying on the three samples to provide anything other than general limiting dates for the two deposits where he found charcoal. We were able to refine the chronology of these units by interpreting the contexts of archaeological sites embedded within, resting upon, or buried beneath the three principal deposits to bracket or refine their ages.

Our geologic work identifies and describes stratigraphic sections where unit relations are clear. Measurements were made by steel tape or by approximations. Attention was given to, 1) buried soils and unconformities, 2) sediment size/source, 3) paleo-current directions, 4) vertical relief on depositional packages, and 5) cultural features diagnostic of depositional environment and/or age. Sixteen profiles were measured and drawn. We were unable to produce an area geologic map, but instead rely on aerial photos from several locals that show unit relations. We were also unable to produce a north-south longitudinal section along the valley showing the vertical relationship of deposits of Holocene age, and refer the reader instead to Christenson's section that included our entire study area (Christenson 1985, Figure 4: 14-15).

Archaeological methods consisted of surface reconnaissance only, looking at pottery and architecture to provide general age estimates for depositional packages or interpretation of habitation environments and agricultural practices. Ceramic and architectural typologies are based on wares and forms defined in the regional Pecos Classification system which we refer to as "Basketmaker" and "Pueblo" for the three Puebloan phases, followed by the phase numerals and occasionally with modifiers such as "early," "middle," or "late" (Table 1.1). All ceramics were examined in the field and left where found. Buried middens were particularly useful in determining minimum age estimates of depositional units upon which they rested or maximum age estimates for units resting on them. Detrital sherds encased in alluvium provide only minimum ages, so their interpretation is inherently ambiguous except in large assemblages. However, it should be noted that only sherds of PII and Pueblo III (PIII) ages were found in detrital depositional contexts; earlier ceramics of BMIII or PI age were only found on ground surfaces along valley margins or adjacent landforms.

Recognizing that Montezuma Canyon presents an unparalleled opportunity for students in archaeology and geomorphology to take a driving tour of this study area, we have organized much of the information in this report from north-to-south along the canyon, and, in a way, that makes the sections and sites readily accessible for inspection and study by others.

Chapter 2
Depositional Units, Stratigraphy, and Geomorphology

Christenson (1985) proposed a remarkably simple Holocene stratigraphy for Montezuma Canyon that we were able to confirm and extend. The two units of Christenson's stratigraphic framework which are relevant to Puebloan occupation are his Units 1 and 2, which he identified at a type locality above Pearson Canyon (Location MC7 in Figure 1.2; Figure 2.1). He was able to recover three radiocarbon samples at that locality to provide a general temporal framework for the two units (Table 2.1). The units extend the length of the study area although they are more distinctive and discernible in the upper canyon than the middle

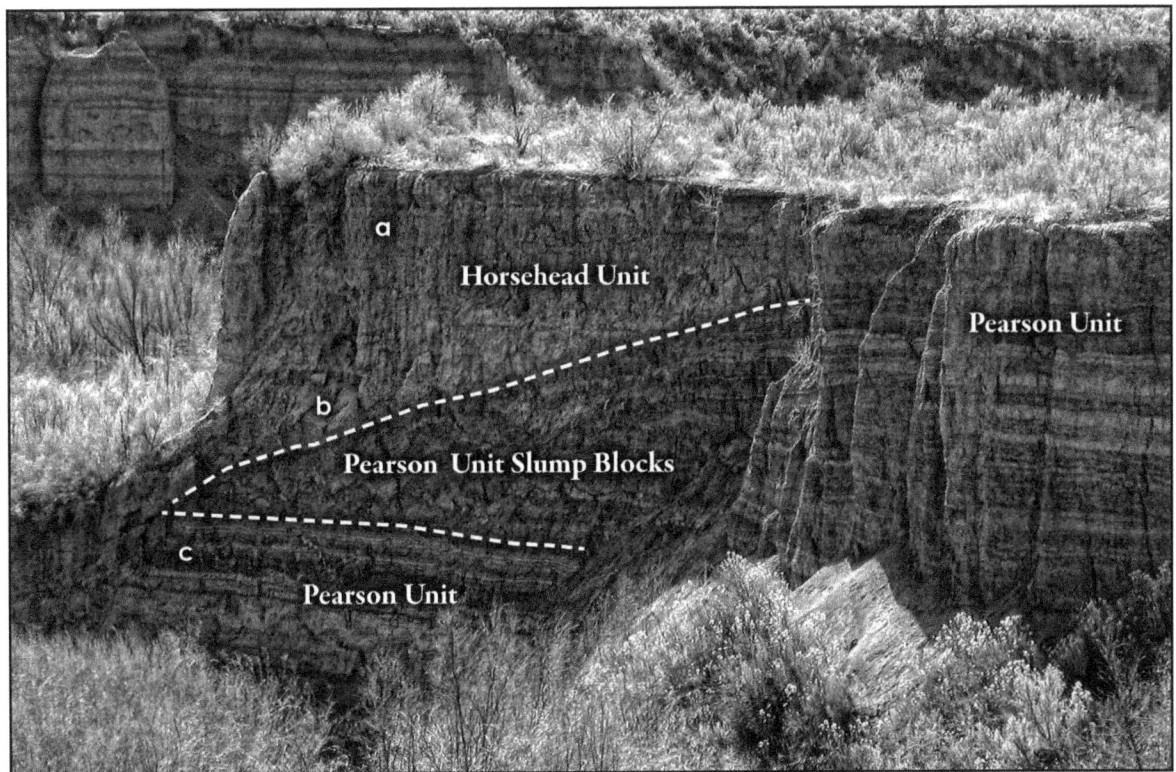

Figure 2.1. Christenson's type site for his Unit 1 and Unit 2, at MC7, separated by an unconformity. Christensen's Unit 1 is the older and his Unit 2 in-fills an arroyo that had cut into it. A large wedge of slump blocks had tumbled from the Unit 1 cutbank and into the arroyo before the Unit 2 deposit began to infill the arroyo. The slump blocks were not apparent in Christenson's 1985 photo. Letters a - c indicate the relative horizontal positions of three radio carbon dates collected by Christenson. This is our Pearson Unit type site. Photograph by Joseph Pachak.

Table 2.1. Radiocarbon Dates from G.E. Christenson's Type Site (Our MC7), 1985, and Keyed to Figure 2.1.

Sample	Radiocarbon age	Calibrated age	Calendar years
a	1410 ± 80 B.P.	1165 – 1515 B.P.	A.D. 435 - 785
b	1430 ± 80 B.P.	1175 – 1530 B.P.	A.D. 420 - 775
c	5130 ± 200 B.P.	5585 – 6345 B.P.	B.C. 4495 - 3635

canyon. We are unable to add any radiocarbon dates to the three Christenson collected, but we can add numerous archaeological sites in contexts that clarify and animate the time intervals involved, and allow us to propose an alternative interpretation for the two radiocarbon dates he applied to his Unit 2. Christenson's descriptions of depositional units are also rather rudimentary, and we are able to better describe them, especially as regards variations and facies within each. We were also able to recognize and define a new unit which Christenson had attributed to overbank flooding and eolian reworking on his Units 1 and 2, but which we see as a separate depositional unit that post-dated the two earlier units.

Christenson identified only three places in the project area where bedrock is exposed in the creek bed (termed "nickpoints"), thus revealing the entire depositional sequence. There are two artificial exposures where modern irrigation diversion structures have exposed bedrock (see Christenson 1985, Figure 4: 14-15). The third is a natural nickpoint just above Three Kiva Pueblo where the creek has exposed a bedrock shelf (the road crosses this exposure). Otherwise, throughout the project area bedrock is concealed and we are only seeing the upper portion of the sediment package.

Predating Christenson's Unit 1 are gravelly terrace deposits, most visible in the middle and lower canyon, which are probably of Pleistocene age and correspond to the shaping of the canyon's walls and bottom. The shaping of the canyon is also quite notable in the cliffed alcoves that define ancient meander loops in the upper canyon reach. Historic-era deposits (Christenson's Unit 3) are also minimally treated in this report; they constitute floodplain and channel deposits built within the pre-1900 arroyo and are named by us and represented in diagrams, but are otherwise left untreated.

So distinctive and widespread are the depositional units that we propose dropping Christenson's numeric labels and giving them names. We name the depositional units for the locality where they were first identified or are particularly representative, applying the names of nearby named features to the units. All are tributary canyons which are the only prominently named geographic features throughout Montezuma Canyon (other than archaeological sites and modern ranches).

PEARSON UNIT

General Description

This older unit of valley fill tends to be the more strongly indurated than younger units due to carbonate or iron-oxide cement, and/or its compact clay matrix. Probably the most distinctive aspect of the Pearson Unit characterizes its upper part, where a nested succession of graded, fine-grained laminated floodplain deposits occurs. Each floodplain deposit commonly has an incipient soil on it, indicting some time-lag allowing for biotic activity before the next flood covered it. In some sections, such deposits alternate with sandier layers and comprise the entire exposed thickness of the unit, in most places 10 m or more. This laminated character persists down-valley in main-channel deposits to greater or lesser degrees at least as far as Tank Canyon, indicating a stable valley floor prone to periodic over-bank floods. The degree of lamination is determined partly by position relative to constricting blockages of the channel by bedrock outcrops and tributary-channel fans.

Upstream of such constrictions, a greater number of floodplain laminations occur, creating places where water flow slowed and sediments settled out. The number of laminations also decreases irregularly downstream (to as few as three) as the canyon widens; side tributaries bring in different source materials and the unit composition changes. Throughout much of the study area where the Pearson Unit is exposed at the surface, its high clay/silt content and induration due to carbonate or iron-oxide cement creates a hardpan.

Toward the base of the unit, sandy beds alternate with clay beds, and may show channels, generally shallow and commonly filled with oxbow-lake fines. Charcoal is locally abundant. At the type area, scores of thin floodplain deposits intercalated with thin sand beds that overlie thicker sand beds intercalated with clay beds to the base of the exposure. This inter-bedding is indicative of a meandering stream anastamosing across an alluvial plain. An outstanding exposure of the unit is visible from the road at study site MC15, below Pearson Canyon, which shows the relationship of floodplain and channel deposits in this unit, including a large in-filled oxbow in cross-section (Figure 2.2). The oxbow would have formed when a meander loop was cut off as the main channel found a new course to the right of the profile. A backwater pond, the oxbow subsequently filled with fine sediments.

Facies

The Pearson Unit's main-channel character occupied the entire valley floor forming a floodplain deposit. Most of the exposures are comprised primarily of overbank flood deposits, which are comprised of compact laminated layers of sands, silts, and clays.

In two places, we were able to find exposures that reveal main channel facies. One is the oxbow illustrated in Figure 2.2, which shows a crescent-shaped channel profile in-filled by fine sediments. The other (MC 14) is at the lower end of Montezuma Village just down canyon from Dodge Canyon (described below). This exposure reveals an active channel that had migrated across the older floodplain, creating an unconformity, with the new channel forming a thick bed of alternating channel profiles in-filled by sands and small gravels interlaced by laminated horizontal flood deposits. This deposit in turn is capped by a compact sand layer that may represent the final channel as it migrated across the Pearson Unit surface.

At the canyon margins where coarse colluvial deposits have accumulated and at alluvial fans sourced by side tributaries, these deposits may be older or younger than the Pearson Unit, but we are unable to demonstrate facies changes of these types for the unit.

Figure 2.2. Pearson Unit exposure at MC15 just downstream from Pearson Canyon, with horizontal flooplain deposits to the right and in-filled oxbow channel to the left. The rightward dip of the surface and underlying horizons is due to compaction of sediments. Photograph by Eric Force.

Age

Only one radiocarbon date for the Pearson Unit was obtained by Christenson (1985), near the base of a laminated exposure some 8 m deep in the deposit, but of unknown relationship to the entirety of the unit since it's base is not exposed at the type locale (Table 2, Sample c). The calibrated date of 5585 to 6345 B.P. would place the accumulation of the visible Pearson Unit from mid-Holocene time and later. It is, of course, likely that the scores of floodplain deposits overlying Christenson's dated horizon represent a hundred years or more, but we do not have any dates for the rate of accumulation nor for the culmination of the Pearson Unit deposition. We did not find any anthropogenic remains embedded within the unit at this locale nor at any of the dozen or so other exposures which we visited throughout the project area, but we did find Puebloan archaeological remains of all ages resting upon its surface at a number of places. We, therefore, interpret the unit as pre-Puebloan in age. This unit thus provides a window into the archaic period paleo-environment in Montezuma Canyon.

Fluvial Paleo-environment

The upper part of the Pearson Unit records a valley filled as far as its geomorphic controls allowed, flooding its floor rather than cutting deep channels, and possibly outpacing the growth of tributary fans. Where fans do occur, such as near the type site at Pearson Canyon or at Bradford Canyon, flood laminae suggest a shallow gradient and quiet aquatic environment, bordering on lacustrine at times.

HORSEHEAD UNIT

With the Pearson Unit in place and filling the valley from wall-to-wall, there followed an episode of erosion that deeply incised the Montezuma Canyon alluvial plain and left an arroyo-scarred landscape reminiscent of today. Such processes are generally diachronous, with incision beginning down-valley and progressing up-valley as the arroyo cuts headward. We do not speculate here on its cause. The arroyo channels remained open for some time allowing for slump blocks to fall from Pearson Unit cutbanks and accumulate on the margins of the arroyos (see Figure 2.1). There followed a period of infilling, which is present at Christenson's type locality, but is best represented by an exposure we found at Horsehead Canyon (MC13), which we propose as the name for this unit (Figure 2.3).

General Description

The Horsehead Unit began to fill the deep, steep-walled arroyos that had cut into the Pearson Unit in a fluvial process very distinct from that which had prevailed before. The Horsehead Unit at the type site consists of interbedded channel deposits with lower gravels and upper sands with fluvial dune-type cross-bedding. At Christenson's type site (MC7), the Horsehead Unit graded to the top of the older unit obscuring their contact, although in other areas (discussed below), it forms an inset terrace below the higher Pearson Unit surface. Some parts of the study area permit map-view reconstruction of the patterns formed by the channels cut by the incision event and partially filled by the Horsehead Unit (for example, see aerial images presented in discussions on MC8, MC10 and MC15 below).

The Horsehead Unit shows considerable variation; indeed its variation is a distinguishing characteristic. Channel fills are evident in most good exposures, whereas floodplain deposits are subordinate in sharp contrast to the Pearson Unit. Channel fills range from clay plugs (lower energy or backwaters) to coarse gravels (higher energy or main channels). Deposits are generally sandy with cross-beds and commonly gravelly, predominantly with pea-gravels. Cementation is generally weak consisting of carbonate (or locally calcium sulfate) and minor iron-oxide, and grains are generally framework-supported as the clay matrix is minor. Clast imbrication (the stacking of grains by water action) is common but shows a wide spread of transport directions, suggesting main channel course migration. Charcoal is common.

High relief at the base of the Horsehead Unit is common where the base is exposed. For example, at Horsehead Canyon (MC13), potholes 5 m deep were eroded into the Pearson Unit and filled by sands and gravels. Such erosion produced detrital mudballs, which, locally, are characteristic of the Horsehead Unit but absent in the Pearson Unit, an indicator that they are derivative of the clays so abundant in the older deposit. Another distinctive inclusion unique to the Horsehead Unit is Puebloan

Figure 2.3. The Horsehead Unit type site. The Horsehead Unit rests unconformably on the older Pearson Unit. It consists of indurated sands and gravels deposited during multiple episodes of channel cut-and-fill episodes. Photograph by Eric Force.

cultural materials, including detrital sherds, arranged rock features (agricultural?), and in a few cases architectural remains.

Christenson (1985) proposed a division into units 2A and 2B to separate deposits below and above an unconformity that he thought might be continuous through much of the valley. He did not describe the two subunits nor give thicknesses. We are not inclined to correlate the many erosional features we saw within the Horsehead Unit with subunits, although, we originally considered that an uppermost deposit of looser sand might be representative of his 2B. However, upon further inspection we are inclined to designate it a separate, younger unit. We, therefore, view the Horsehead Unit as composite in nature and representative of an active fluvial system but lacking in extensive unconformities that would allow separation into subunits.

Facies

The Horsehead Unit shows a greater variation of depositional types than the Pearson Unit, due in part to greater development of alluvial fans derived from side tributaries impinging on the main channel. The main-channel deposits are generally coarse-grained. Typical are cross-bedded sands with pea-gravel layers grading on valley margins into bouldery colluvial deposits where channels impinged within a few tens of meters of steep bedrock slopes.

Tributary alluvial fan deposits are prominent and, where graded to main-channel deposits, can extend well into the Montezuma Canyon valley floor as well as reaching up their tributary drainages. Their deposits tend to be better-sorted and less-indurated sands than the main-channel deposits, and where tabular pebbles are present, they show clast imbrications indicating direction of flow out of the tributaries. The mouths of Bradford Canyon, Tank Canyon, and Coalbed Canyon provide particularly good examples of fan/floodplain interfaces.

Age

Christenson (1985:20) reported two radiocarbon dates from the Horsehead Unit at his type locality (Table 2.1, Samples a and b). One date was from near the base of the unit, the other from near the top (see Figure 2.1). These calibrated dates of 1175 to 1530 B.P. and 1165 to 1515 B.P. (A.D. 420 to 775 and A.D. 435- to 785; average median about A.D. 606), respectively, are virtually contemporaneous, implying that infilling of the arroyo was nearly instantaneous, with the entire unit forming within a few decades during BMIII time. While we allow that this may be a valid interpretation, per the discussion in the methodology section about the uncertainties of interpreting radiocarbon dates and the utility of using sediment embedded charcoal to provide exact dates for deposition, we do not believe they can be reliably used to date the depositional event nor rate of accumulation. We interpret the two radiocarbon dates to indicate that the Horsehead Unit infilling of the old arroyo as a maximum age having occurred sometime after the median A.D. 606 date as interpreted by Christenson. In contrast, we found a PI habitation site (likely dating to the mid-to-late 800s) down valley near Three Kiva Pueblo that had been built on the Pearson Unit surface, but was later truncated by the pre-Horsehead Unit incision event (see discussion of MC6 below), which means it was in place on the Pearson Unit surface prior to the incision event. Furthermore, we found a number of study sites down-canyon where mid-PII sherds and features were found embedded within the Horsehead Unit sediments. Since incision-aggregation is diachronous, with the process developing first down canyon and progressively moving up canyon, we would expect older deposits to occur down canyon and become progressively younger up canyon. However, if we accept that the arroyo in the Pearson Unit had filled in late BMIII time at MC7 as interpreted by Christenson, how do we explain the younger materials in Horsehead Unit fill down valley?

We interpret Christenson's dates not as marking the actual infilling of the arroyo, but as maximum bracketing dates derived from old wood and indicating only that the accumulation of the Horsehead Unit occurred after Basketmaker time. This is understood with the suite of younger Puebloan archaeological materials down canyon clarifying the timing of the incision event as more likely occurring during late PI or early PII time, with the infilling of the arroyos underway by mid-PII time.

Fluvial Paleo-environment

At the type locality, the channel filled by the Horsehead Unit records a pronounced re-activation of the fluvial landscape of Montezuma Canyon. The Pearson valley floor that had been incised by deep, steep-walled arroyos began to in-fill with inter-bedded sands and gravels. Within this "inner valley," the main channel was actively cutting some sections headward while depositing the cuttings down valley, where they would accumulate forming channel fans, which, in turn, would be cut by later incision channels. Overall, the process was accretive and, through time, the cut-and-fill process began to fill the old arroyo with sediments, in some places to a level below the old Pearson Unit valley surface and forming an inset terrace (as at MC8, MC10 and MC15) and, in some places, filling it to the same level (as at MC7). This unit relationship and fluvial process defined the principal landscape shaping process during PII and PIII time.

Nancy Patterson Unit

The Nancy Patterson Unit was identified as a separate unit late in our study when it became apparent that an upper layer of much looser sand occurred discontinuously throughout the study area. In fact, its discontinuity — we found it in only seven of the 17 study sites we examined — is one of its defining attributes. We name it for a study site in the wash just west of Nancy Patterson Village where its association with the underlying Horsehead units is pronounced (Figure 2.4).

General Description

Where present, the Nancy Patterson Unit consists of particularly loose sand which includes both fluvial and eolian deposits, but lacks the cementation of the Horsehead Unit. As such, it occurs as a thin mantle of loose inter-bedded sand sitting atop the more indurated lower units. In some locations, such as MC2 near Nancy Patterson Village, it rests atop the Horsehead Unit, while at others, such as just above Coalbed Village, historic erosion has removed the Horsehead Unit entirely from its mid-valley position and we find the Nancy Patterson Unit resting directly atop the Pearson Unit. At Nancy Patterson Village (MC2) we found it represented as a point bar in a narrow channel, but in most cases it is represented as a mantle of loose laminated sand deposited by overbank flooding from a filled channel, and subsequently reworked by eolian action.

Facies

The unit, where present, generally consists of medium-to-fine bedded sands as channel deposits, or more commonly as overbank floods. These loose deposits have subsequently been reworked by the wind into low dunes. The unit also occurs on the tributary fans at Tank Canyon and at Bradford Canyon indicating sheetwash discharging across the fan surface from a full tributary channel, probably grading to a similar depositional environment that was occurring on the main valley floor.

Age

The Nancy Patterson Unit lacks the cementation of the Horsehead Unit, which is consistent with it being younger. In five locations, it was found to overlie middens or structures from mid-PII to mid-PIII age. In one location (MC3) we found it overlying a midden of late PII/early PIII age with a historic rubble mound associated with a wagon road resting on its surface, bracketing its age between late Puebloan time and the Euroamerican historic period. Our conclusion is that the Nancy Patterson Unit

Figure 2.4. The Nancy Patterson Unit type site near where the trail to Nancy Patterson Village crosses Montezuma Creek at MC2. The Nancy Patterson Unit rests on a more compact Horsehead Unit deposit. Photograph by Deanne Matheny.

began to form no earlier than late PII time but more probably after PIII time. Since we do not find evidence of re-entrenchment of the canyon floor until the historic event around 1900, we presume the main channel remained actively aggrading, though discontinuous, for the centuries following Puebloan abandonment of the canyon.

Fluvial Paleo-environment

In the places where the Horsehead Unit did not succeed in totally in-filling the arroyos left from the early entrenchment episode we do not see evidence of the Nancy Patterson Unit, indicating the fluvial "complex response" process in the main channel remained consistent or active. It is only where the Horsehead Unit filled the entrenchment that local sheet wash spread onto the Horsehead and/or Pearson surfaces, depositing a thin mantle of sands that were later reworked by wind. As such, the presence of the Nancy Patterson Unit deposited discontinuously throughout the project area is an indicator of a system in equilibrium with active cutting and filling where stream flow was contained within the inner-gorge, but forming the Nancy Patterson Unit in areas where the old arroyo banks were over-topped and sheet flow dispersed out across the valley floor.

COALBED UNIT (CHRISTENSON'S UNIT 3)

The Horsehead Unit and, where discontinuously deposited, the Nancy Patterson Unit seem to have remained a stable depositional environment for a number of centuries following Puebloan times; we do not see evidence of substantial aggradation nor do we see any evidence for the onset of new erosion, that is, until the historic period. During the late 1800s into the early 1900s, the onset of erosion that occurred throughout much of the southwest affected Montezuma Canyon. Beginning down valley, the erosional front moved up canyon and cut deep channels into the valley floor, flushing much sediment out of the drainage and leaving a deeply incised valley floor. But as the fluvial process continued, channel fans began to form down valley and build headward infilling the bottom of the deep arroyo. When truncated by down cutting by the active stream, this younger deposit forms the terraces within the historic arroyo. We named this unit for the vehicle pull-out area at Coalbed Village (MC8), an area with which many readers will be familiar.

STRATIGRAPHY AND GEOMORPHOLOGY

Our investigations show that the Pearson Unit was likely accreted for several millennia by a meandering stream channel that regularly overflowed its banks, laying flood deposits that came to fill the canyon from wall to wall. This was probably the situation throughout the project area, but is most pronounced in the upper canyon where the unit is best preserved and exposed. This valley floor was incised by a severe erosional event which occurred during early Puebloan time, most likely late during the PI period or early in PII. In the upper canyon, several cross-sections of the boundary are exposed and show steep to moderately sloping walls cut into typical Pearson Unit flood deposits, which were subsequently buried almost to the top by typical Horsehead Unit channel deposits.

In the middle canyon, such cross-sections are rare, and except for a few instances where we could locate valley margin remnants of Pearson Unit deposits (such as MC3 and MC17), the Pearson Unit valley fill appears to have been largely removed during the incision event and subsequently filled by Horsehead Unit channel fill or over-topped by Nancy Patterson Unit sheet flow. In the upper canyon, the upper surface of the Horsehead Unit is commonly lower than the Pearson Unit; that is, it forms an inset terrace and the main-valley fill is of two levels at such places. Where the Horsehead Unit is inset into the Pearson Unit, the boundary faces the modern drainage (as at MC8 and MC10). It is the geomorphic relationship of these two units that formed a significant component of the Puebloan habitation landscape. Late during the Puebloan occupation of Montezuma Canyon, but more likely subsequent to it, the Nancy Patterson Unit was deposited discontinuously throughout the project area, but may have been of limited consequence to the Puebloan farmers. Historic erosion has exposed these geomorphic relationships, which are depicted in Figure 2.5.

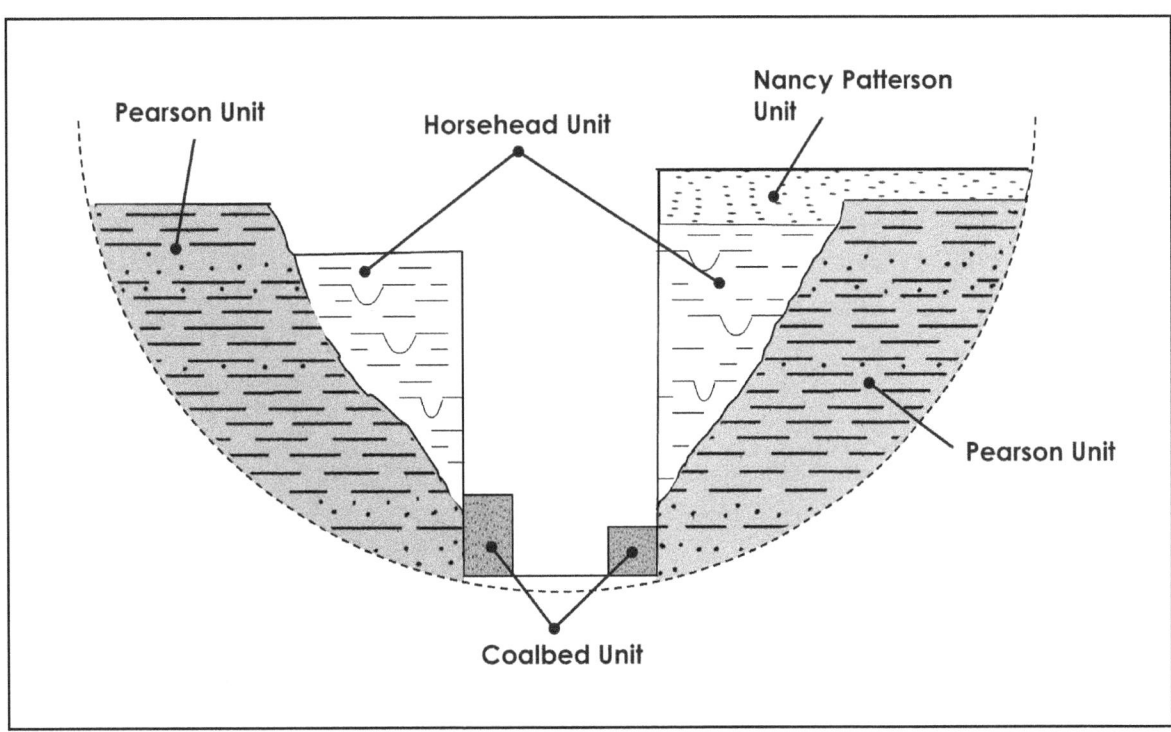

Figure 2.5. Schematic cross-section showing the stratigraphic relationships of the four alluvial units in Montezuma Canyon. The left side of the cross section shows the relationship when the Horsehead Unit is an inset into the Pearson Unit, the Nancy Patterson Unit is absent or not differentiated, and the Coalbed Unit is inset below the two older units but conceals their basal contact (as at MC10). The right side depicts the situation where the Nancy Patterson Unit tops both of the older units, and the younger Coalbed Unit does not conceal the basal contact between them (as at MC8).

Chapter 3
Archaeological Relationships

Our understanding of the Puebloan adaptation to this landscape is made clear at 11 localities spread throughout the study area — eight in the upper canyon and three in the middle canyon. At each locality we found good exposures of the units, and archaeological materials in contexts that provide chronological controls as well as reveal ways in which Puebloan settlers responded to the topography, soils, and hydrologic environments. In fact, so distinctive are these associations in several locations that it is possible to recognize the actual agricultural fields utilized by Puebloan farmers. From north to south, we describe each locality and organize them in a way that they can be revisited and studied by our readers and future students. Most are accessible or visible from the county road.

Panorama Point Locality (MC15)

This study locale is just downstream from the mouth of Pearson Canyon at the canyon bend immediately above Dodge Canyon and Montezuma Village and provides an unmatched 180 degree panoramic view of the depositional history. This section of canyon is strongly influenced by the large alluvial fan emanating from Pearson Canyon. As mentioned in the preceding section, above Pearson Canyon, the Horsehead Unit graded level with the upper surface of the Pearson Unit (at MC7), but, at this locale, it did not come close to that level, suggesting the Pearson Canyon fan formed a constriction that influenced different depositional processes above and below the canyon mouth. This locale provides one of the best vantages in the canyon for understanding the relationships of three of the four alluvial units (the Nancy Patterson Unit is absent). This is because following the prehistoric erosion event, the Horsehead Unit only filled to a level about five meters below the upper surface of the Pearson Unit, and when that valley floor was subsequently cut by the historic erosion event, the Horsehead Unit was isolated as an inset terrace below the Pearson with a Coalbed Unit terrace forming below it (Figure 3.1)

The channels carved by the two erosion events are evident in Figure 3.1. When the pre-Horsehead erosion event happened it may have bisected the valley differentially leaving a plug of Pearson material in the middle of the valley (see Figure 3.2), although, what now appears as a plug may have been part of the northside Pearson Unit terrace, a plug of which was subsequently isolated by the historic erosion event. Regardless, this highly dissected valley floor was utilized by Puebloan people who built residences or field houses on the stable landforms of the Pearson terrace on both sides of the valley, and perhaps even on the isolated island of stable land in the middle of the valley. We observed evidence of four Puebloan residences or activity areas, two with mid-

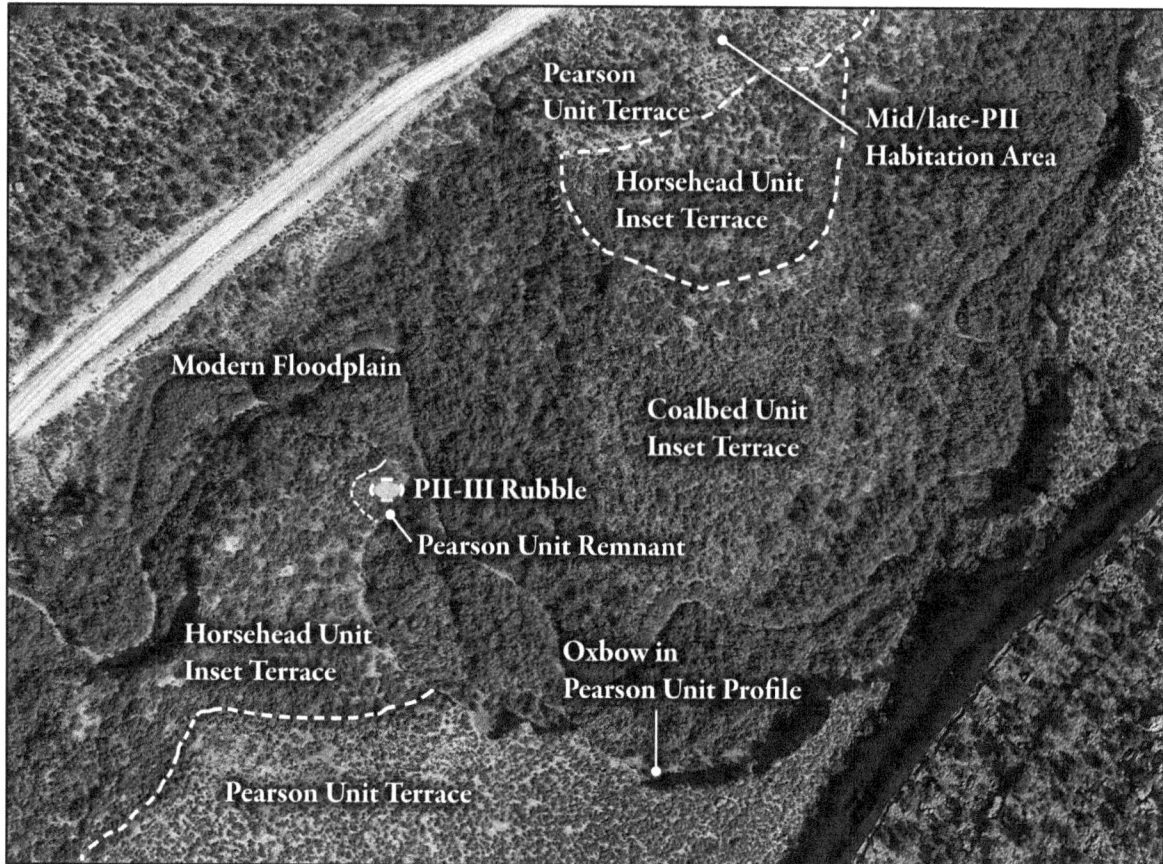

Figure 3.1. Aerial view of MC 15 study site just below Pearson Canyon. Puebloan habitation sites and activity areas are found on the Pearson terrace, with the Horsehead terrace inset about 4 to 5 m below it, and the Coalbed terrace inset below that. Image taken from Google Earth, 2017.

to-late PII ceramics and two which we could not access, but which have masonry rubble mounds indicative of PII or PIII architectural practice. The fluvially active inner-valley floor would have concentrated water, sediments, and nutrients — optimum conditions for floodwater farming. For the Puebloan farmers to have farmed on the Pearson Unit surface (and utilize the entire valley floor), they would have needed to lift the water some 4 to 5 m from the entrenched stream level. This would have required a long irrigation ditch, which would have been a daunting engineering challenge; one which historic farmers attempted and failed to accomplish (an abandoned historic ditch follows the contour of the Pearson Unit cutbank through the locality). We interpret the optimum agricultural ground in this stretch of canyon as occurring on the aggrading Horsehead Unit surface, which, because the Pearson terrace edge is preserved on both sides of the valley, allows us to measure a 75 to 125 m wide valley bottom where water could have been diverted onto fields. With canyon widths from bedrock-to-bedrock ranging from 250 to 350 m wide (300 median) and the Pearson Unit occupying much of the canyon, only about one-third of the valley floor would have been near the water table and suitable for agriculture.

Figure 3.2. Isolated plug of the Pearson Unit at MC15, with a Horsehead Unit inset terrace to the left and the modern cutbank to the right (see Figure 3.1 for plan view). The distinct units are evident in the strong horizontal bedding of the Pearson Unit as opposed to the Horsehead Unit's uniform coloring and lack of strong horizontal bedding. Note the rubble from a Puebloan masonry room atop the knob. We could not access the knob but based on the masonry rubble presume the site to be Pueblo II-III in age. Photograph by Eric Force.

Montezuma Village, Upper (MC5) and Lower (MC14) Localities

Just down canyon from the Panorama Point locality, the depositional unit relationships are less pronounced. At the time of our field work, we were limited in our ability to access most of the private property in Montezuma Village, so were able to look at only two profiles — one on the north side of the village and one near the south end. They provide similar perspectives.

The alluvial fan emanating from Dodge Canyon shapes this canyon bottom, with its west-to-east slope forcing the Montezuma Creek channel to the east side of the valley. At the Bull property at the north end of Montezuma Village (MC5), we were able to find two PII habitation sites that illuminate the ancient topography. On the west side of the modern drainage we found a PII habitation site exposed on the modern ground surface that was built directly on a layer of a meter or so of indurated sand. This hard sand layer sits in turn on a deep exposure that is clearly comprised of the Pearson Unit. This same indurated sand also comprises the modern surface at the lower end of Montezuma Village (MC14 discussed below), and may represent the sandy fill of a late channel on the surface of the Pearson Unit alluvial plain.

On the east side of the modern drainage, at MC5, we identified a terrace of the Pearson Unit but it was heavily colluviated because of its proximity to the east canyon wall. We found a PII unit house on that terrace surface. The presence of these two PII sites on either side of the modern drainage indicates that the Puebloan-age channel occupied more or less the same space as the modern drainage. The edges of the Pearson terrace are not intact here so exact measurement of the Horsehead-aged valley bottom width is not possible, although, the two archaeological sites lie about 150 m apart, indicating the PII-aged floodplain would have been less than that, but of a similar width to what we saw at the Panorama Point locality.

At study site, MC 14 on the southern end of Montezuma Village, we found good exposures on public land (Figure 3.3). On the east side of the valley, near the base of the slope, we found a colluvially buried midden of PII age exposed in the cutbank near the base of the bedrock canyon slope, similar to what we found at MC5. On the west side of the modern drainage, we found an excellent exposure with the two facies of Pearson Unit valley fill (flooplain and main channel). A middle PII midden was found resting on the surface unit of indurated main channel sand, similar to what we saw at MC5.

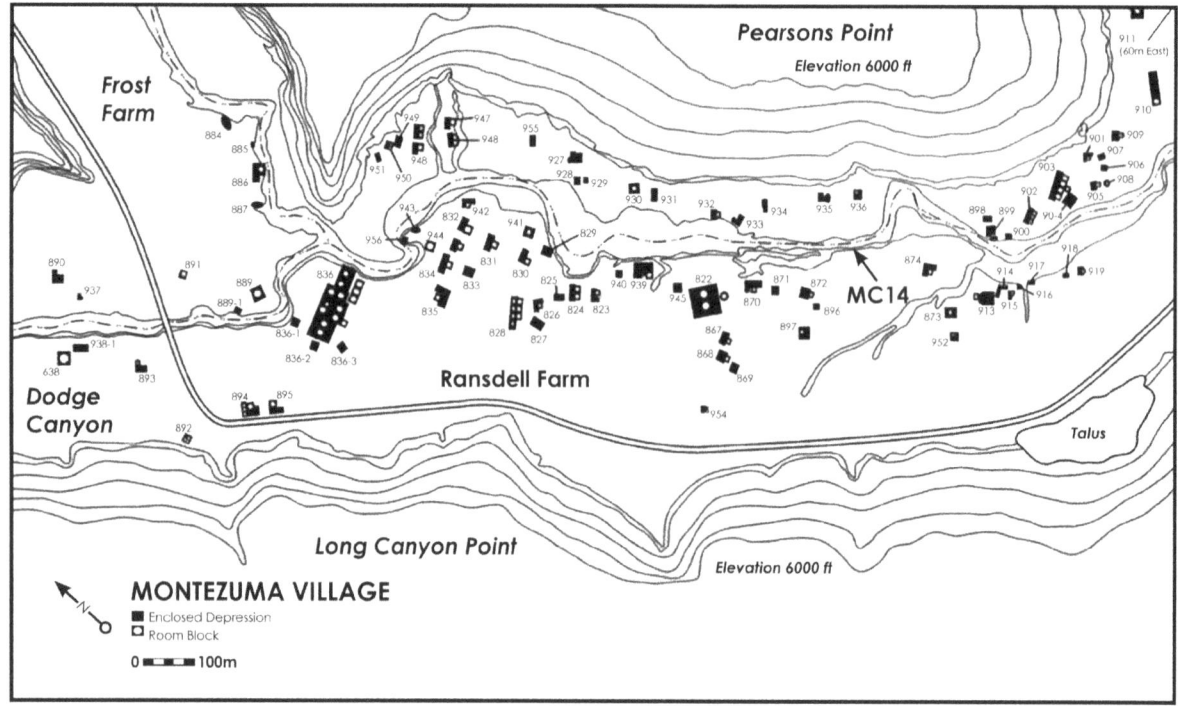

Figure 3.3. Map of Montezuma Village, showing location of the MC14 study site. Puebloan sites and artifacts are found on both sides of the drainage, which date from Basketmaker III through Pueblo III, and indicate the main channel was constrained to more or less its present course throughout Puebloan time (map adapted from Baer 2003, which was adapted from Matheny 1962).

Here too, west-side alluvial fan slopes force the drainage to the east side of the valley, and the presence of multiple Puebloan sites on the Pearson Unit surface on both sides of the modern drainage indicate that the channel has been constrained more or less to its present course since before Puebloan time.

In Figure 3.3, the six sites depicted on the eastern bank are separated from the dozen or so sites on the west bank by about 80 m, indicating the channel flow during Horsehead times was constrained within a narrower area than what we observed up valley at MC7. Subsequent to our field work, Howell was able to pay a brief visit to the Ransdell property in the middle portion of Montezuma Village where he was able to clarify these observations and identify the contact between the Pearson Unit and the Horsehead Unit, which consists of a narrow inset terrace about one meter below the Pearson Unit surface. This subtle contact is actually visible on the Google Earth images on both sides of the valley and the distance between them measures about 90 m, similar to what we see at MC14.

With the Horsehead Unit beginning to in-fill the arroyo during Puebloan time, a good agricultural base was available in a ribbon stretching along the entire length of Montezuma Village. Combined with water discharged from the perennial tributary stream in Dodge Canyon, and given the distribution of Puebloan sites across the entire canyon floor, it is reasonable to assume that much of the Montezuma Canyon bottomland was suitable for agricultural use here throughout Puebloan time, with the entire canyon bottom offering prime conditions.

Horsehead Canyon Locality (MC13)

This study locale was recognized by Frank Lesure for the shared relationship of alluvial deposits and Puebloan architecture. We were able to identify the site depicted in Lesure's sketch and add to his depositional descriptions as well as clarify archaeological contexts. Horsehead Canyon is an ideal locale for understanding the relationship of a tributary alluvial fan to the main canyon alluvial plain. Figure 3.4 shows the locations of observations discussed here.

The distribution of channel cross-sections within the Horsehead Unit indicates a channel that was cutting and filling within the same deposit, carving potholes as much as 5 m deep into the Pearson Unit that were filled by sands and gravels. Sediments discharged from the Horsehead Canyon alluvial fan aggraded into and interlaced with this main valley deposit, and the area appears to have been favorable for agriculture with water influencing sediment deposits flowing from two directions. From the county road, we were able to observe, from a distance (on private property), two places in the modern cutbank of the Horsehead Canyon fan where sandstone blocks are visible protruding from the sediments, in both cases, several meters down in the profile and well out on the floodplain and away from any possible colluvial influence. In one locality, two blocks sit in close proximity to each other and are encased in channel sediments, an indication they had intentionally been placed in an actively aggrading channel, perhaps as water diversion features (Figure 3.5). A PIII habitation (cover image) was built on the

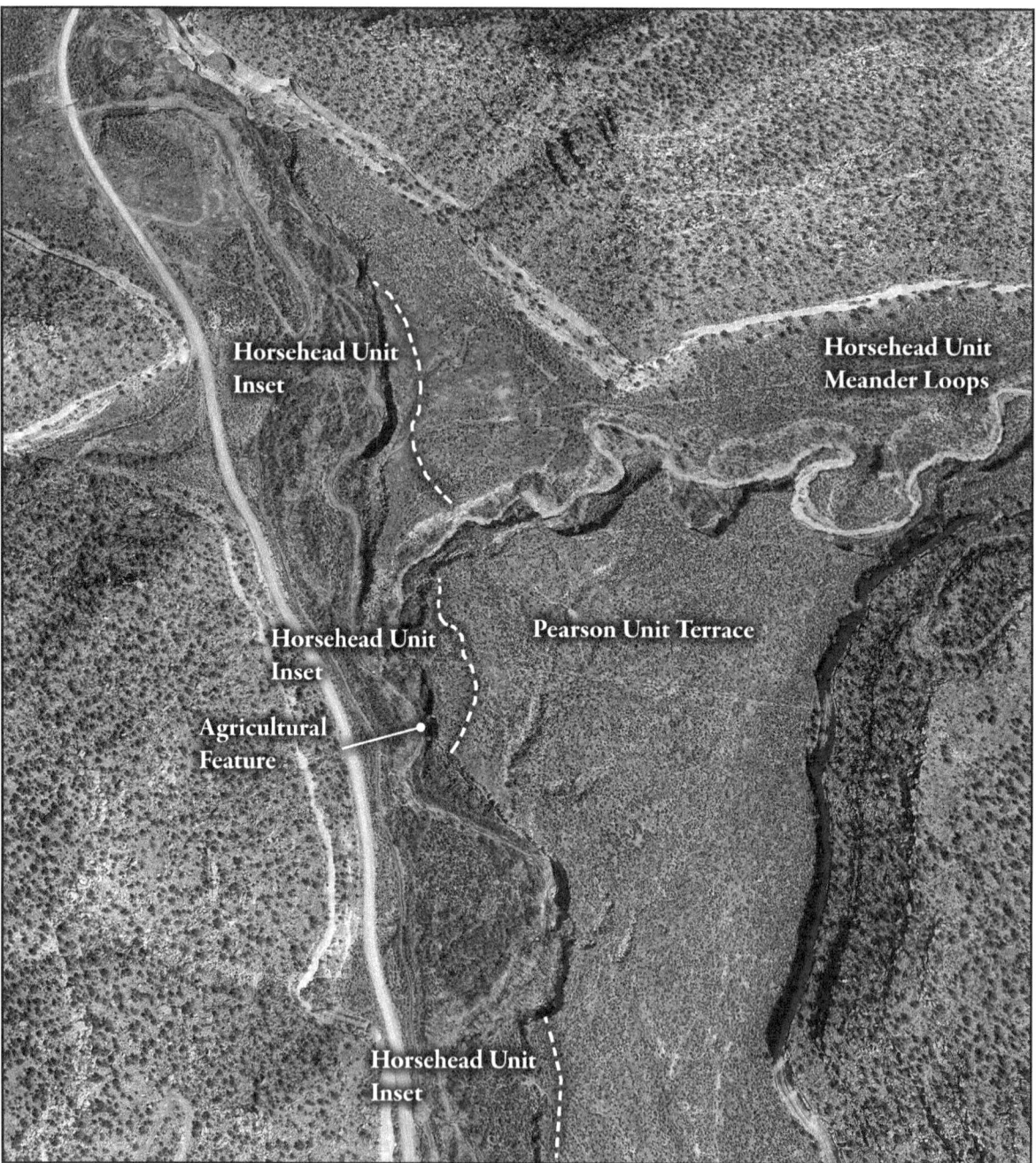

Figure 3.4. The confluence of Horsehead Canyon and Montezuma Canyon at MC13. Deposits of Horsehead Unit age are visible as inset terraces within the Pearson Unit, with younger channels up canyon carved into the older deposit. Image taken from Google Earth, 2017.

Horsehead-age alluvial fan surface in a position adjacent to the tributary stream, and together with the sediment encased rock features suggest the Horsehead Canyon locality was a favorable place for agriculture.

Figure 3.5. Two sandstone blocks in profile, which were placed in proximity to each other in an actively aggrading channel and may have been part of an agricultural feature. Channel cross-sections visible in the profile suggest channels braiding across the floodplain. Photograph by Eric Force.

COALBED VILLAGE LOCALITY (MC8/MC11)

We discuss here a stretch of canyon from about a half kilometer above Coalbed Village to a kilometer below it as there are three places where depositional/cultural relationships are revealed (Figure 3.6). This locale provides information relevant to all aspects of our understanding of the canyon's Holocene geomorphic history, the influence of tributary alluvial fans, and the Puebloan settlement adaptation to a varied landscape. This is a complex package of landforms and human history, and our limited time at the site allows only a tentative interpretation. All of the features discussed here are on public land and readily accessible from the county road.

The defining topography of the Coalbed Canyon locale is that the principal archaeological site is built on the top, sides, and basal platform of a bedrock island mesita that was isolated from the surrounding canyon walls in the distant past by the migrating meanders of Montezuma Creek as well as the significant outflow from Coalbed Canyon. As a consequence, the island mesita lies at

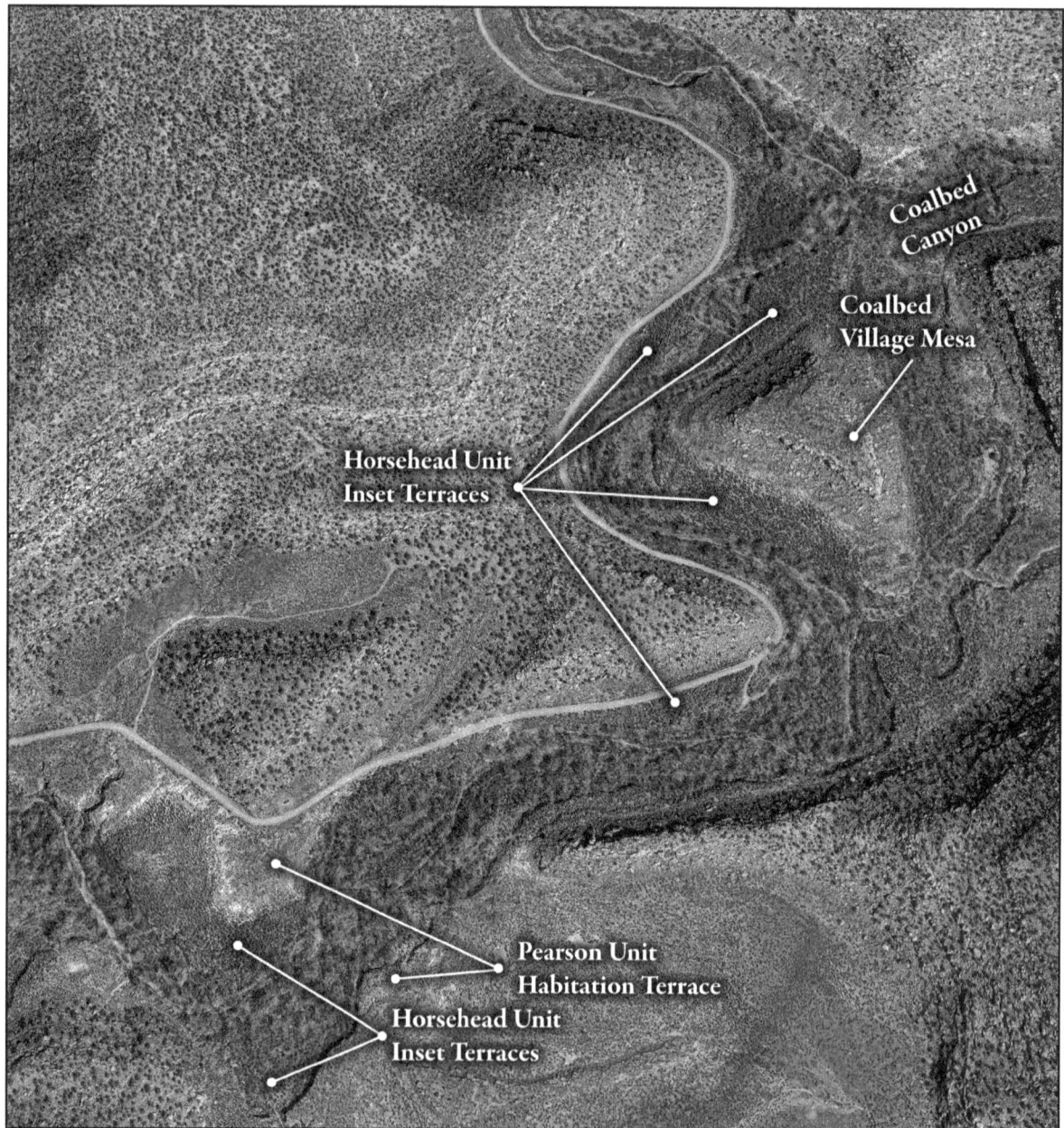

Figure 3.6. Overview of the Coalbed Canyon vicinity (MC11 at top of image, MC8 at bottom). The Pearson Unit forms a basal layer supporting colluvium along the bedrock canyon margins as well as around the Coalbed Village mesa. Horsehead Unit inset terraces are visible as areas of darker vegetation, consisting primarily of greasewood. Note in the lower left of the image the distinct edges of the arroyo that cut into the Pearson Unit valley floor prior to Horsehead Unit deposition. Image taken from Google Earth, 2017.

the center-point of a complex set of fluvial influences flowing in from two directions, and the story revealed above the canyon junction is quite distinct than the story below.

Study location MC11 is located in the canyon bend a half kilometer above the mouth of Coalbed Canyon. At this location, the relationship of the Pearson and Horsehead units has been removed by modern erosion, but the Pearson Unit has been preserved as a distinct terrace along the valley margin. At this location and for several kilometers up canyon from it, the Nancy Patterson Unit is present as a mantle of sand directly overlying the Pearson Unit (as noted at MC12). In contrast, the Nancy Patterson Unit is absent at Coalbed Village and for some distance down canyon. This situation may be a consequence of the large alluvial fan that emanated from Coalbed Canyon and effectively wrapped around the mesita, creating a constriction which slowed water flow.

When the prehistoric incision event happened in Puebloan time, it carved an arroyo into the canyon bottom but left a remnant of the Pearson Unit intact as a terrace at the base of the bedrock mesita. As the Horsehead Unit accumulated in the old channel, it did not overtop the Pearson Unit terrace and instead formed as an inset terrace that eventually rose to a level a meter or so below the older surface.

The narrow Pearson Unit terrace became an area of intense habitation where multiple generations of Puebloans arranged their architectural units side-by-side but oriented in relation to the terrace edge and the aggrading Horsehead valley floor below it (Figure 3.7). In some instances, Puebloan middens and building rubble draped the terrace slope below habitation areas. Detrital corrugated body sherds found in Horsehead Unit depositional contexts indicate a minimum age of PII for this deposition. What is now a greasewood flat would have existed as an aggrading floodplain at least throughout PII to PIII time and would have provided significant agricultural potential to local inhabitants. Had the architecture been in place prior to the erosion event that carved the Pearson Unit cutbank, we would expect the architecture to be truncated and the middens absent, but they are not, indicating they were built or deposited onto the already eroded landscape.

This locale also provides some perspectives on the relationship of the alluvial fan emanating from Coalbed Canyon with the main valley sediment package. At the south end of the village site near the parking area and on the east bank, we observed 2 to 3 m of the Horsehead Unit fan facies emanating from Coalbed Canyon resting atop the Horsehead Unit valley fill. In the fan deposit, we found several detrital sherds including corrugated wares, indicating an actively aggrading alluvial fan during at least PII time, similar to materials observed elsewhere where alluvial fan growth outpaced aggradation on the main valley floor.

Study section MC8 is located just down canyon from Coalbed Village, where the influence of the tributary fan is less pronounced. The geomorphic relationship between the Pearson and Horsehead Units is quite distinct here (see Figure 3.6).

At this study locale, the Pearson Unit terrace was never over-ridden by the Horsehead Unit so exists as a distinct upper terrace on both sides of the valley bottom, with the younger unit inset about one meter below the Pearson surface. The arroyo banks left from the incision event are particularly distinct on the north side of the canyon. On the east bank of the modern creek, there is a unit pueblo consisting of a roomblock and pit structure with a ceramic assemblage that dates to the early to middle PII period. It was built on the margin of the Pearson Unit terrace near the aggrading Horsehead Unit floodplain. Because we have intact Pearson terrace edges on both sides of the canyon floor, we are able to measure the width of the Horsehead floodplain, from 125 to as much as 175 m wide. This floodplain, similar to what we saw up canyon (MC15), would have provided favorable conditions for floodwater farming.

Figure 3.7. Rubble mounds of lower Coalbed Village. The Pearson Unit forms a terrace at the base of the mesa where much of Coalbed Village is situated. It was heavily occupied throughout Puebloan time as indicated by a continuous band of rubble in this image. The Horsehead Unit alluvial plain formed as a surface several meters below the habitation area. Photograph by Deanne Matheny.

RATTLESNAKE SITE LOCALITY (MC10)

This locale just below the confluence of Devil's Canyon provides one of the best examples in the study area of geomorphic unit associations and the Puebloan adaptation to the landscape. It is an extension of the landscape described for the lower Coalbed Canyon landscape (MC8), with two distinct terraces — a higher Pearson terrace with the Horsehead inset terrace about a meter below it. An unusual feature of this landscape is what appear to be several low hillocks that occupy the central part of the valley floor. Upon inspection, we found the hillocks to be relics of the Pearson Unit that had been isolated from the main valley floor during the prehistoric incision event. This association is quite visible in Figure 3.8.

The hillocks apparently offered an attractive location for habitation during middle PII time, as a Puebloan farmstead consisting of a masonry roomblock and kiva was built there (we named it the Rattlesnake Site for a snake we observed on the site). The farmstead was occupied long enough to accumulate a substantial midden. Modern erosion has

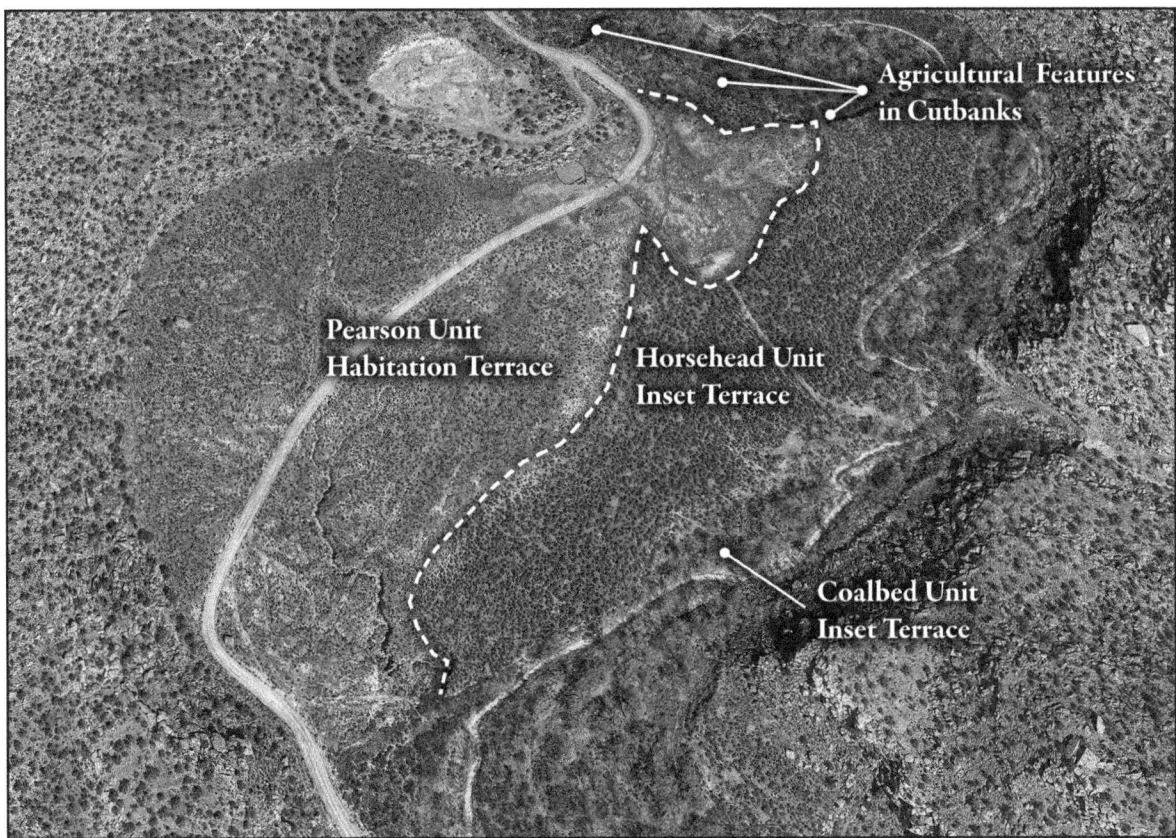

Figure 3.8. Aerial view of the Rattlesnake Site vicinity (MC10), showing the sharp contrast between the Pearson Unit and the Horsehead Unit. Alignments of sandstone blocks, perhaps agricultural features, are common in the Horsehead unit profiles. Image taken from Google Earth, 2017.

exposed a striking profile on the north edge of the site which reveals the association of the habitation area with its contemporary floodplain (Figure 3.9). The cutbank exposure reveals three linear rock features embedded in Horsehead Unit context — horizontal alignments of unshaped sandstone blocks that likely represent check dams or water diversion features. The rock features are variously located 2 to 3 m below the Horsehead ground surface and are encapsulated in alluvially-deposited sands and pea gravels, indicating that the features were constructed on small channels of an actively aggrading floodplain. Ceramics found in association with the rock features and as detritus elsewhere in the surrounding deposit are, at minimum, PII age or younger, in the case of white wares or corrugated body sherds, and when types could be identified they appear to be contemporaneous with the Rattlesnake Site midden materials.

In addition, along the contact between the Pearson and Horsehead terraces south of the habitation area, we found several artifact scatters consisting of localized soil stains, lithics, and ceramics which are also contemporaneous with the Rattlesnake habitation site. We predicted the presence of these artifact scatters after we had recognized the relationship of the Pearson/Horsehead terraces in this locality and after we

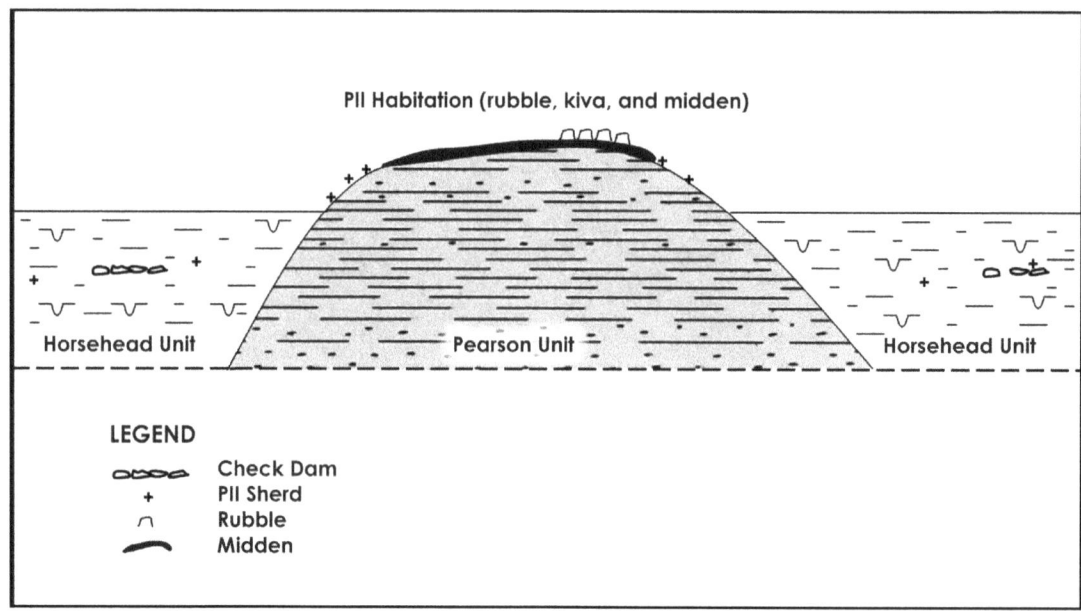

Figure 3.9. Profile showing the relationship of the Pearson and Horsehead units at the Rattlesnake Site, and associated Pueblo II archaeological remains.

had found the buried water diversion features. We interpret them to represent field houses or field camps, situated on the Pearson terrace cutbank and in close proximity to agricultural fields that would have been located nearby on the aggrading Horsehead floodplain.

BRADFORD CANYON ALLUVIAL FAN LOCALITY (MC9)

This study site provides excellent exposures for understanding the relationship of the main valley floodplain and side canyon alluvial fans, and the Puebloan adaptation to the alluvial fan environment. Where Bradford Canyon issues its sediment discharge into Montezuma Canyon, it creates a broad alluvial fan that spreads out on both sides of a bedrock mesita that occupies the canyon mouth. This mesita hosted human occupations that spanned the Puebloan period, with a large PIII rubble mound the most conspicuous.

There are two good exposures visible near the point where the county road crosses the dry streambed issuing from Bradford Canyon. East of the road, the cutbank on the south side of the stream offers an exposure where a calcium carbonate-rich Pearson Unit valley fill is over-ridden by a Horsehead Unit alluvial fan (Figure 3.10). Here, as elsewhere in the study area, we do not find evidence of tributary canyon fans in the Pearson Unit depositional context, prompting us to conclude that, in this locality throughout Pearson Unit time, main valley deposition overwhelmed input from the side canyons, whereas during Horsehead Unit time, sediment discharge from side canyons overwhelmed main valley deposition and pushed the alluvial fans well out onto the canyon bottom.

This created an environment attractive to Puebloan farmers. For some distance along the south-side cutbank of the Bradford Canyon

Figure 3.10. Facies profile at the road crossing of the Bradford Canyon stream (MC9). The Horsehead Unit alluvial fan rests directly on the Pearson Unit valley fill. Photograph by Wayne Howell.

streambed, we found places where sandstone blocks indicative of Puebloan architecture or field features, were either protruding from or eroded out of the cutbank, in some cases several meters down in the profile. We also found several corrugated body sherds indicating deposition of at least as old as PII age.

Just south of the Bradford Canyon streambed crossing, we found exposed in the road cut on both sides of the county road an in-situ early PIII site with a crude masonry wall of unshaped blocks and a midden (Figure 3.11). These cultural remains occur about 2 m below modern ground surface and suggest a fairly extensive activity area. The cultural deposit is sitting on an indurated Horsehead Unit surface and is covered with an eolian altered deposit of mixed gravelly sand, indicating that overbank floods had covered the site and those deposits had subsequently been altered by wind. This uppermost part of the unit may be a fan-facies of the Nancy Patterson Unit.

THREE KIVA PUEBLO LOCALITY (MC6)

This study site is located just downstream from a juncture where a bedrock shelf is

Figure 3.11. A Pueblo III-age roomblock exposed in the east cut bank of the county road at Bradford Canyon. Photograph by Winston Hurst.

exposed, creating a natural nickpoint. It is a 100 m-long chute where the slope of the main drainage channel steepens sharply and the creek bed flows across bedrock (Figure 3.12). We found that, above the nickpoint, the depositional package had thinned considerably and virtually pinched out at the head of the chute. However, below the nickpoint, the Pearson Unit/Horsehead Unit relationship so prevalent up-canyon is present, albeit less distinct, as the Pearson Unit characteristics had changed; there were fewer of the finely laminated flood deposits near the top of the unit. Vertical and horizontal exposures of the two main depositional units and the presence of Puebloan sites in association with them provide a primary clue to the timing of the prehistoric erosion event that bisected the Pearson Unit.

It appears that the Pearson Unit floodplain was deposited below the nickpoint much as it had been up canyon — overbank floods deposited layers of sands, clays, and laminated flood deposits across the valley floor, lapping up against the valley edges where colluvium was integrated into the facies (the colluvial component is readily visible in the cutbank where the stream runs against the country road down canyon from Three Kiva Pueblo). Subsequently, erosion cut a channel into the Pearson floodplain, though not to bedrock as revealed in a profile visible to the west of Three Kiva Pueblo (Figure 3.13). The subsequent Horsehead Unit alluviation filled the arroyo and over-topped the eroded Pearson Unit surface in the central part of the valley, but did not overtop it on the valley edges where a low inset terrace

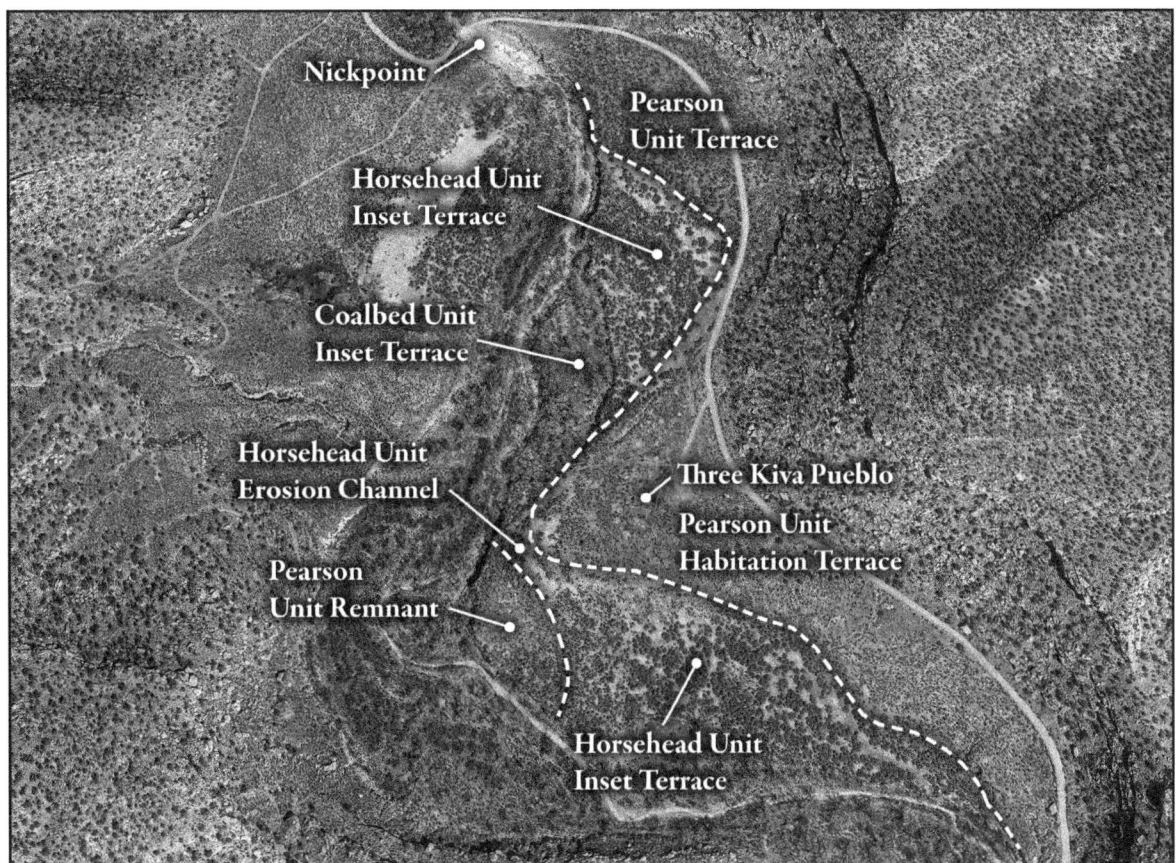

Figure 3.12. Aerial view of the Three Kiva Pueblo vicinity, showing the nickpoint, unit associations, and the location of Three Kiva Pueblo (MC6). The dashed white line shows the contacts between the Pearson and Horsehead units. The erosion channel depicted in the lower left center of the image is shown in Figure 3.13. Image taken from Google Earth, 2017.

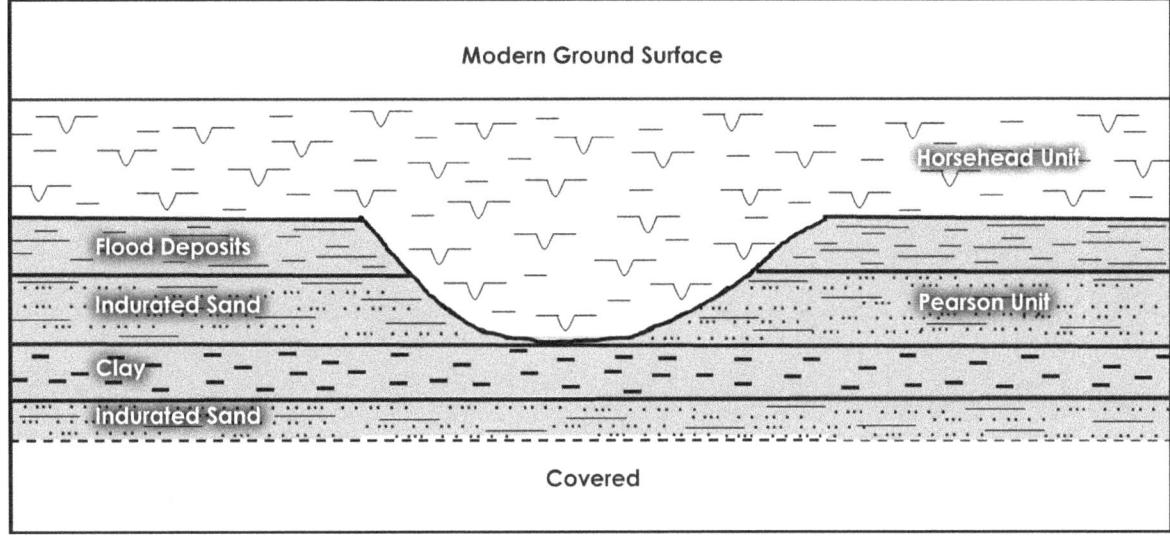

Figure 3.13. Profile showing the relationship of the Pearson and Horsehead units in the middle canyon floor as exposed in the cutbank just west of Three Kiva Pueblo (profile represents about 4 m of deposition).

remained. Former arroyo banks formed by the erosion event are evident in cross-section here as they are up-valley (see Figure 3.12).

Excavations carried out by the BYU field school (D. Miller 1974) demonstrated that Three Kiva Pueblo had a substantial PI occupation in buried contexts, with substantial PII and PIII overbuilding. The village was situated well above the floodplain on what may be a sediment-capped bedrock shelf, so its relationship to the floodplain is not evident. However, another small PI unit was constructed about 75 m southeast of Three Kiva Pueblo and is visible today on the modern ground surface. It had been constructed on the Pearson Unit surface and is revealing for what is not there. Today, the site consists of a single block of three or four contiguous surface rooms as indicated by several rows of upright slabs. This block of rooms sits on the very edge of the Pearson/Horsehead contact. If we presume that its layout was typical of the unit pueblo architectural convention of the time, we would expect to see a pithouse to the south of the room block and a midden beyond that to the southeast. However, those features are absent; instead, that space is a meter or so lower in elevation and occupied by the Horsehead Unit inset terrace (Figure 3.14). This association suggests that the PI unit house had been built on the northern margin of the valley floor and occupied prior to the prehistoric erosion event. That incision event subsequently truncated the site, removing the midden and pit house but leaving the surface rooms intact. Therefore, the timing of the erosion event in the Three

Figure 3.14. A Pueblo I roomblock is visible as rubble and several rows of upright sandstone slabs. The slope break in the right center of the image is the contact between the Pearson and Horsehead Units. Photograph by Joseph Packak.

Kiva locality would have occurred during or after PI time. Testing at this site could refine the date of the occupation and help bracket the timing of erosion.

A late PII to early PIII habitation site was later built on the Pearson Unit terrace edge on the south side of the valley opposite the larger pueblo, and adds to our understanding of the Puebloan agricultural situation, as the location of this habitation site on the Pearson terrace edge gives us a general dimension of the aggrading Horsehead floodplain during Puebloan occupation — on the order of 130 m — similar to what we observed up canyon.

TANK CANYON LOCALITY (MC4)

In the Tank Canyon vicinity, modern erosion has eliminated most of the Pearson and Horsehead units so we could not find good exposures to study in the main valley. However, good exposures along the axis of the alluvial fan emanating from Tank Canyon have revealed two archaeological sites in depositional context, both of which are informative in dating the Horsehead Unit fan-facies depositional history as well as illuminating Puebloan land use practices. Both archaeological sites are exposed in the eastern cutbank of the tributary fan arroyo, and both date to the middle PII period, although, they are not contemporaneous.

The older site is a buried midden (Bond 1984). The cultural deposit extends for about 8 m along the exposure and is rich in ash, charcoal and artifacts. It averages about 50 cm thick and is currently buried about 90 cm below the modern ground surface. It rests on indurated sand and is covered by a non-indurated upper sand whose surface has been altered by eolian action and likely represents a Nancy Patterson-aged fan facies. Ceramics indicate occupation during the early 1000s. If the midden had been situated to the southeast of the habitation features — the common practice — by implication, the architecture associated with the midden has been entirely eliminated by modern erosion.

The younger site is located about 40 m to the north (and up-gradient) of the buried midden. Modern erosion has bisected the site exposing a surface roomblock and pitstructure (kiva) in profile (Figure 3.15). There is scant surface evidence of the site, consisting of only a few sandstone cobbles which are out of context on the otherwise eolian sandy surface. The pitstructure had been excavated into compact sand. A masonry bench was constructed around the interior perimeter and a layer of red clay was placed on the floor. A masonry-lined wall, which included upright slabs and associated pilasters, was constructed above the bench. With a roof in place, the level of the pitstructure would have been at about the same level as the floor of the surface room. After the pitstructure was abandoned, it began to fill periodically with water, and subsequent decay of the walls sent rubble tumbling into the pit. The thickness of the ponded sediments beneath the wall tumble — about a meter thick — indicates that the first ponding event(s) brought in substantial sediments, likely sheet wash from overbank foods originating from Tank Canyon. Such sheet floods would have also shallowly buried the surface roomblock.

We found no evidence to indicate whether the site was occupied at the time of flooding. Examination of the ceramics indicates the site was occupied in the mid-1000s, a generation or so later than a nearby buried midden site. The presence of these two near-contemporaneous sites on the Tank Canyon alluvial fan demonstrate that it provided a desirable habitation environment

Figure 3.15. A Pueblo II unit pueblo has been bisected by modern erosion. Rubble from a surface room lies near the surface to the left and is buried by about 20 to 30 cm of bedded sand, likely the Nancy Patterson Unit. A kiva is visible as vertical pilaster and bench columns, and the floor is marked by a horizontal slab to the right. The kiva was filled by multiple episodes of overbank flooding emanating from the Tank Canyon drainage. Photograph by Joseph Pachak.

for a number of decades during the first half of the eleventh century. Although we cannot say whether flooding was a cause for abandonment, habitation there was not without risk, as sheet wash from overbank floods emanating from Tank Canyon eventually inundated both sites.

Farther down gradient on the fan, near the tributaries' juncture with the main canyon, we found that the fan deposit graded into the main channel deposit, the lower part of the fan facies in particular becoming more clayey. We also found a main-channel oxbow that had filled with carbonate cemented, horizontally-bedded, and laminated sand lenses (each about 1 cm thick) derived from the main channel. In this area, where the fan graded into the main valley deposit, we continued to find evidence of the distinct units of the Horsehead Unit fan-facies — a compact indurated lower unit overlain by a less indurated surface unit about 2 m thick. On the contact between the two, we found corrugated body sherds sticking out of the cutbank, an indicator that the Tank Canyon fan was aggrading onto the main valley fill during, or later than, PII time. This indicates that both the Tank Canyon fan and the main valley were aggrading environments with the oxbow channel an indication that the main valley had a stream coursing across its surface. This scenario — a filled valley with a stream and aggrading tributary fan — made the confluence of Montezuma and Tank canyons a desirable place for agriculture and helps explain the multi-unit, multi-component PI through PIII villages built on the bedrock ridge nearby.

Because of the prevalence of private property below Tank Canyon, we had very few opportunities to examine exposures, and have only three study sites to discuss for the middle canyon. But they tell a coherent story — the almost complete absence of the Pearson Unit and the prevalence of the Horsehead Unit.

CAVE CANYON LOCALITY (MC16/MC17)

In the vicinity of the mouth of Cave Canyon, we were able to find two localities on public land that illuminate landscape history. At the road intersection where Cave Canyon enters Montezuma Canyon, Eric Force was able to locate, just within the road right-of-way, an exposure in the Cave Canyon arroyo where a coursed stone masonry wall was exposed several meters below the modern surface (MC16). Later examination of the profile indicated the masonry was the wall of a kiva with a masonry roomblock exposed in the cutbank nearby. The roomblock was capped by about a meter of loose sand. Ceramics show the habitation to be of middle PIII age. The kiva had been excavated into a Horsehead Unit deposit and both the room block and kiva had been covered by about one meter of loose interbedded sands of Nancy Patterson Unit age, the later unit having pooled into the kiva depression sometime after it had been abandoned. Just down canyon from the PIII unit, we also found some building rubble and artifacts eroding out of the cutbank. Ceramics indicate a late PII to early PIII occupation, likely a generation or so earlier than the structure just up canyon. This presents a scenario very similar to what we noted at Tank Canyon and Bradford Canyon, where the tributary canyon fans provided a favorable habitation environment during PII and into PIII times, but one prone to overbank flooding.

Subsequent to the main field work, we had noted on Google Earth imagery a kilometer or so up Montezuma Canyon from the Cave Canyon road intersection what appeared to be a contact between two depositional units (MC17). Upon examination, we confirmed that it was the contact between the Pearson Unit with a slightly inset terrace (<1 m) of Horsehead Unit age. The Nancy Patterson Unit was absent which explains why we were able to discern the contact remotely. At the edge of the contact, we found on the Pearson Unit surface a habitation of mid-PII age, and nearby another habitation of mid-PIII age. We were unable to examine the Horsehead Unit surface here because it was on private property. This locality is significant because it was the only good exposure we confirmed of the Pearson Unit surface in the middle canyon.

BIGHORN SITE LOCALITY (MC3)

Several kilometers south of the Cave Canyon intersection, a section of public land (MC3) provided us good exposures where we were able to identify the four depositional units so prevalent in the upper canyon (Figure 3.16). We also found archaeological materials exposed in depositional context to provide for relative age controls. At a place we call the Bighorn Site (for a petroglyph found nearby), we were able to examine exposures along both cutbanks of Montezuma Creek. Both lengthy exposures are comprised primarily of the Horsehead Unit deposits similar to what we saw up-canyon — an indurated calcareous sand with some charcoal and coaly layers, which has been channeled and in-filled by subsequent episodes of erosion and deposition. This slightly indurated unit was subsequently capped by an upper unit of <.5 m of loosely bedded sands of the Nancy Patterson Unit, which is prevalent throughout this stretch.

Where the modern arroyo has cut against the bedrock ridge on the north side of the valley, we found one small pocket where a remnant of the Pearson Unit has been preserved under colluvium at the base of the ridge. This exposure includes a contact between the Pearson Unit and the Horsehead Unit in a context remarkably similar to what we observed at MC7, albeit smaller in scale, with the exposure only about 2 m high as opposed to the 10 m we observed up canyon. The Horsehead component of this terrace extends several hundred meters east from the county road.

During late PII time, a unit pueblo was built on this 30 m-wide terrace. The site includes a rubble mound indicating a coursed masonry room block with a separate but adjoining rectangular arrangement of upright rocks suggesting a remodeling episode. To the south of the roomblock, a depression indicates a kiva. Exposed in the terrace cutbank, five or so meters south of the kiva, there are several nested pit features exposed in profile. These pits had been excavated into an undulating compact sandy ground surface of Horsehead Unit age, but the surface lay about two vertical meters below the level of the adjacent roomblock, suggesting the occupation area sat astride a slightly eroded ground surface on the valley margin.

About 20 m to the east of the architecture, a small arroyo has bisected the terrace and exposed in profile the site's late-PII to early PIII midden. The midden had been deposited on the same undulating Horsehead Unit surface as described above (Figure 3.16). This surface stands out as a distinct contact that slopes from north-to-south, indicating it had been eroded prior to deposition of the midden. Sometime subsequent to the deposition of the midden, overbank floods of the Nancy Patterson Unit encapsulated the terrace edge (but did not reach far enough to cover the nearby rubble mound and kiva). The last episode of cultural activity at the site involved a historic wagon road that had angled down the slope of the nearby bedrock ridge, and at the contact where the road ran onto the valley floor, someone had stacked a pile of sandstone rubble, perhaps as a retaining wall or more likely simply to remove obstructions. This rubble wall clearly sits atop the Nancy Patterson Unit sand. This study site, thus, provides us with the two best bracketing dates we have for the age of the Nancy Patterson Unit. It clearly postdates early PIII time, though by how much we do not know, and was in place and forming the valley floor prior to early historic times when a wagon road was built through the site area.

These geomorphic and cultural relationships indicate that the erosion event that cut arroyos into the Pearson Unit was so extensive that it largely eliminated that unit from the valley in this area. The Horsehead Unit that subsequently filled the valley was periodically channelized and in-filled by fluvial action out on the valley floor, while, on the valley margin away from the active floodplain, the landform provided a stable enough surface for Puebloans to build upon. It remained stable for sufficient time — at least a generation or so — for a midden to accrue

Figure 3.16. An arroyo has bisected a compound terrace in the Bighorn Site area, exposing a late Pueblo II to early Pueblo III midden in depositional context (MC3). The sloping, eroded nature of the Horsehead surface is marked by a dashed line. The Nancy Patterson Unit clearly post-dates the midden, and the historic rubble wall associated with a wagon road rests atop the Nancy Patterson Unit, bracketing its age. Photograph by Kyle Bauman.

and for some architectural remodeling to occur. This hardened surface was subsequently over-topped by bedded sands deposited by fluvial action, likely overbank floods from a floodplain that had continued to aggrade out on the main valley floor subsequent to the Puebloan abandonment.

NANCY PATTERSON VILLAGE LOCALITY (MC1/MC2)

We were able to examine two exposures in the vicinity of Nancy Patterson Village near the confluence of Cross Canyon with Montezuma Canyon. We did not find any archaeological associations at either of the study sites, but both sites deserve mention here for other reasons. The MC2 locality is accessed along the main foot trail to Nancy Patterson Village, with the best exposures on the eastern cutbank and north of the trail. This is the type site for the Nancy Patterson Unit, which here consists of about 2 m of loose, laminated pebbly sand which shows cross-bedding indicative of fluvial action. This deposit is devoid of cultural materials and postdates Puebloan occupation. However, the older Horsehead Unit in this locality likely formed an active depositional environment available to the farmers of the nearby Nancy Patterson Village, and the deposit contains a rich array of organic materials that should provide an enhanced view into the depositional history and natural environment during much of Puebloan time.

Here, the Horsehead Unit is comprised of an upper indurated clayey sand, locally about a meter thick, overlying bedded layers of clays and pebbly sands, commonly incised by deep channels that were filled in some places with laminated clay lenses and in others with indurated laminated pebbly sand representing more active channels. Armored mud balls — derived from the erosion of the Pearson Unit clays up-drainage — are common throughout the deposit. Another common inclusion is charcoal abundant in places (Figure 3.17).

We did not find any architecture, constructed features or artifacts in any of the exposures we examined. However, in looking closely at the charred wood, we noted that much of it consists of branches and stems of small plant species (about the size of greasewood, salt bush, or willow). Such short-lived species,

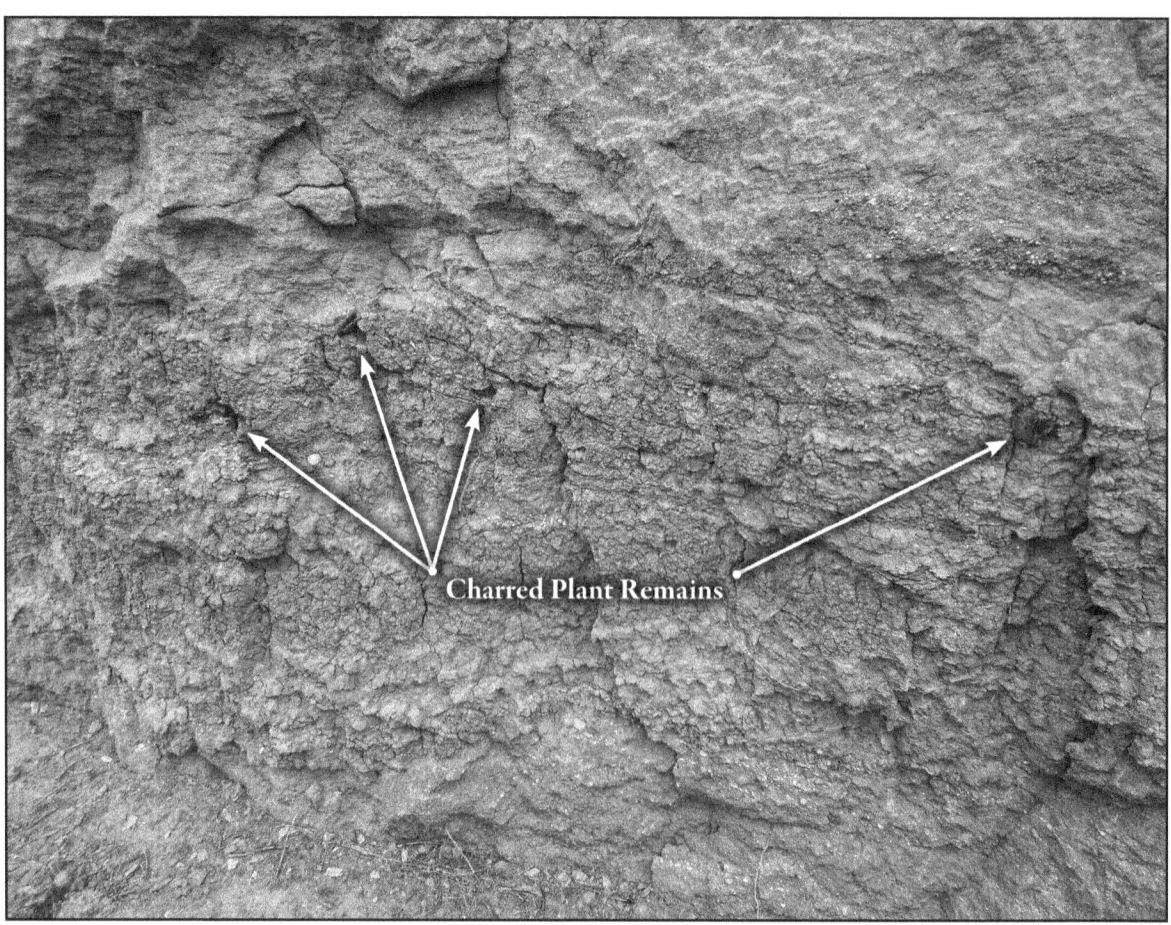

Figure 3.17. Pieces of charred wood are common in inter-bedded clay and sand channel fills, as are fragments of organic materials encased in clay. Preservation of the charcoal is excellent and field examination suggests identification as to species should be possible. Photograph by Deanne Matheny.

and small, relatively delicate items that would not persist long in the natural environment or survive water transport, could address the "old wood" dating problem discussed above, and provide more precise dating for the depositional history of the Horsehead Unit in this part of the canyon. The cause of the burning of course could be natural, but it could also represent cultural activity such as burning of vegetation to clear land for fields. Pollen analysis of the surrounding sediments could address that issue.

Study site MC1 is located on the south bank of Cross Creek at the point where the road crosses the creek. Although the exposure presents a similar looking depositional history — older clayey unit/younger sandier unit — there is significantly more clay content in both units than we found in the main Montezuma Canyon, possibly because of greater clay input emanating from Cross Canyon sources. This results in more indurated deposits, particularly for the younger unit. If the older unit represents the Horsehead Unit, we did not see any channel cutting/infilling in the profile, nor armored mud balls so indicative of the unit in Montezuma Canyon. It could also be the Cross Canyon equivalent of the Pearson Unit. Due to the fact that we did not have time to examine deposits up Cross Canyon, we can only speculate that different controls and processes there shaped the depositional history. We, therefore, leave Cross Canyon with a question mark, but a fitting southern boundary for our study area.

Summary of Depositional and Archaeological Relations

Archaeological materials found in both surface and buried contexts provide chronological controls for determining the relative ages of sedimentary units and serve as indications of past Puebloan land use practices. We did not find any cultural materials embedded within the Pearson Unit, but throughout the study area, we found Puebloan sites and artifacts of all Puebloan phases resting on its surface. We observed what look to be BMIII cists on the Pearson Unit surface, but in valley margin settings at Coalbed Village and on a fan surface at Bradford Canyon, and PI sites in similar valley margin settings at Coalbed Village and at Three Kiva Pueblo.

In contrast, habitation sites of PII and PIII age are common on the Pearson Unit surface in the upper canyon and, in the one place, we found it as a surface in the middle canyon. Many of these sites were situated in relation to the edges of the prehistoric arroyo, indicting it was in place when the habitations were established and they were placed there intentionally. As the Horsehead floodplain began to fill the prehistoric arroyo, detrital artifacts, at least as old as PII age, began to be incorporated within the floodplain deposit and on tributary canyon alluvial fans, in several places associated with likely agricultural features. By mid-to-late PII time, parts of the Horsehead floodplain had stabilized to the point that it also appears to have become an attractive settlement area, at least in the middle canyon where we find architecture and middens from that time period and later resting on its surface in valley margin settings, or imbedded within Horsehead-age tributary alluvial fans throughout the study area. This combination of Pearson and Horsehead surfaces provided the valley bottom platform for farmsteads through late PII and PIII time.

The Nancy Patterson Unit formed in places where the shingling effect of the cut-and-fill process had filled the old Pearson arroyo with floodplain deposits, causing overbank floods to spread out onto a broader area of Horsehead and/or Pearson surfaces. Elsewhere, if contained within the confines of the Pearson arroyo, the cut-and-fill process continued to play out on the active floodplain, precluding over-bank floods and formation of the Nancy Patterson Unit. In several places, we found the Nancy Patterson Unit as a mantle covering sites of PII and PIII age, and although it may have begun to form in the middle canyon as early as late-PII time, it more uniformly appears to post-date Puebloan occupation of the canyon.

Chapter 4
The Puebloan Agricultural Landscape

Our reconstruction of the geomorphic history of Montezuma Canyon demonstrates two distinct fluvial environments and two distinct canyon bottom topographies that likely existed during Puebloan time. These two depositional environments and topographies are separated by a sharp division — an erosion event that appears to separate the BMIII and PI period occupations from later PII and PIII times. This separation is indicated in part by distinct site distributions as recorded for the upper and middle canyons by this study but more thoroughly by BYU researchers.

The Pearson Unit was a suitable habitation surface during all periods of Puebloan occupation. The survey carried out by Ray Matheny at Montezuma Village illustrates this pattern (Matheny 1962; see Figure 3.3). Matheny documented 91 distinct structural clusters with artifact scatters along a one mile stretch of canyon where the perennial stream at Dodge Canyon flows into Montezuma Canyon. Although most of the visible architectural remains date to the PII and PIII periods, these later sites likely obscure older architecture as roughly half of the sites also contained sherds dating to the PI period with some perhaps extending into BMIII time. Occupation sites tend to concentrate on both margins of the modern stream course, which was where the prehistoric arroyo was cut into the Pearson Unit surface, and, hence, also likely the location of the stream channel during BMIII and PI time (any migrations of the stream channel would have destroyed these sites).

In contrast, the survey carried out by Petrus de Haan in the middle canyon (de Haan 1972), from Tank Canyon to the confluence with Cross Canyon that documented 56 sites, found that the canyon floor was also a preferred habitation environment (43 percent of all sites), but, in that stretch, he did not find any sites specifically attributable to BMIII or PI age. This finding is likely explained by our observation that in much of the middle canyon the Pearson Unit had been removed by erosion, with the younger Horsehead Unit comprising most of the observable canyon floor, or topped by the later Nancy Patterson Unit. Hence, site age distribution in the middle canyon is not a function of the Puebloan settlement pattern, rather, the age of depositional units — sites of Basketmaker and early Puebloan age would have been removed by erosion prior to deposition of the Horsehead Unit, with the surface of that unit becoming habitable only by mid-PII time. However, the fact that BMIII and PI populations built numerous villages on stable landforms immediately adjacent to the canyon floor attest to their presence throughout the length of the middle canyon.

The Early Puebloan Landscape

The older topography is represented by the Pearson Unit and indicates a filled valley with a stream meandering across its surface, carving and filling oxbows and laying down fine sediments during periodic overbank floods. An excellent profile of this stream channel is visible at the Panorama Point locality (MC15). The overall depositional trend was accretive. The water table was high and stable, even lacustrine at certain times and localities. Where tributary canyons flowed into the main valley, alluvial fans appear to have graded into the main valley floodplain and remained subservient to it.

The first Basketmaker farmers to arrive in the canyon may have found a filled valley with a stream meandering across its surface. Although Ray Matheny found evidence for the presence of BMIII people out on the canyon floor, for the most part, village sites from that period are commonly found above the valley floor (de Hahn 1975; Matheny 1962) on low ridges just above the alluvial bottomlands, such as the Basketmaker component at Cave Canyon Village (Christensen 1980; Harmon 1977, 1979; Nielsen 1978), on canyon benches, or in saddles several tens of meters above the floodplain, such as Hidden Village (Montoya 2008). Since the floodplain would have been prone to over-bank flooding (at least seasonally), the best sites for settlement would have been on the colluvially-influenced valley fringes, on elevated and sloping alluvial fans at the mouths of some tributary canyons, or on bedrock finger ridges and canyon terraces above the floodplain.

A similar settlement strategy may have prevailed during early PI time, as many settlements are found in locations similar to BMIII sites, such as the PI component at Cave Canyon Village (Harmon 1977, 1979; Nielsen 1978; Christensen 1980), on nearby elevated landforms, such as Monument Village (Patterson 1975) and Three Kiva Pueblo (D. Miller 1974), or atop isolated mesas, such as the large PI components at Nancy Patterson Village (Janetski and Hurst 1984; Janetski and Thompson 2012) and at Tank Village and Coalbed Village. A model of this landscape is presented in Figure 4.1.

The Late Puebloan Landscape

The younger topography was shaped by a more energetic fluvial environment. It consisted of a deeply incised valley floor, with vertical walled arroyos very similar to the modern topography. Based on the presence of many PII sites built in direct relation to this arroyo, and the inclusion of PII artifacts embedded within the newly forming floodplain deposit, we know that the arroyo was carved before mid-PII time. Following the incision event, the water table fell and the main channel was constrained within the arroyo's confines. In the upper canyon, places where water could be spread onto arable land became constricted to the active floodplain within the confines of arroyo or on the alluvial fans at tributary canyon junctures. In contrast to the modest formation of tributary fans during Pearson times, during Horsehead times, fan formation at many canyon mouths was robust, with fans consistently overtopping valley fill and growing out onto the floodplain.

The complex response created a fluid, up-canyon migrating agricultural environment, with sediments dislodged from shallow headward cutting arroyos being deposited down valley as channel fans. These aggradation nodes, where sediments (and nutrients)

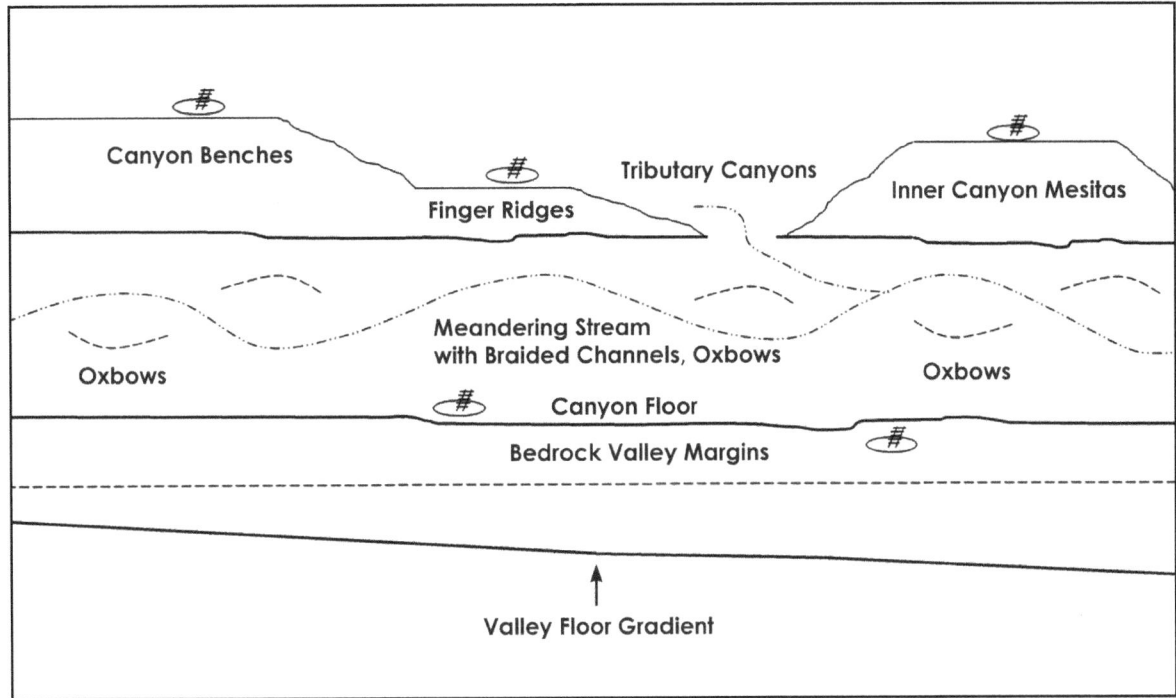

Figure 4.1. A possible cultural landscape during Basketmaker III and Pueblo I times. Some artifacts from those periods are found on the old canyon floor but villages tended to be located above the floodplain on stable valley margins and nearby on tributary fans, atop isolated mesitas and on canyon benches.

deposited and water flow spread out, appear to have created conditions suitable for ak-chin agriculture. In the upper canyon, we find farmsteads and field camps arranged along the Pearson-age arroyo edge and stone agricultural features encapsulated within Horsehead-age channel fan deposits. In the middle canyon, some places on the Horsehead Unit surface became an attractive, though at times perilous, habitation environment as we noted at MC16 (mouth of Cave Canyon) and MC11 (Bighorn site).

As the complex response created incision fronts that towed aggradation nodes up canyon with them, any given stretch of canyon might have been a good place for fields for one generation of farmers, but perhaps not the next, prompting migration within the system. Furthermore, within the system, there was not just one incision front and one aggradation node at any given time, rather multiple. This factor may account for unique site settlement histories; a certain place might have been attractive when the aggradation node was nearby, abandoned when the node moved upstream, then reoccupied when a subsequent node migrated up-valley.

In contrast, tributary canyon fans were places where sediment discharge was confined within a bedrock-constrained valley and formed as "fixed" aggradation nodes. These alluvial fans would have provided agricultural potential on their surfaces and they would have spread favorable conditions out onto the main valley floodplain as they grew. Thus, it is reasonable that the largest and longest inhabited villages occur at the major canyon mouths (e.g. Montezuma Village, Coalbed

Village, Devil Canyon Village, Bradford Canyon Village, Tank Village, Nancy Patterson Village). The fact that some villages — for example, Montezuma Village and Nancy Patterson Village — appear to have been inhabited for the entire span of Puebloan time, attests to the productivity of these canyon junctions for agriculture. Figure 4.2 presents a model of this landscape.

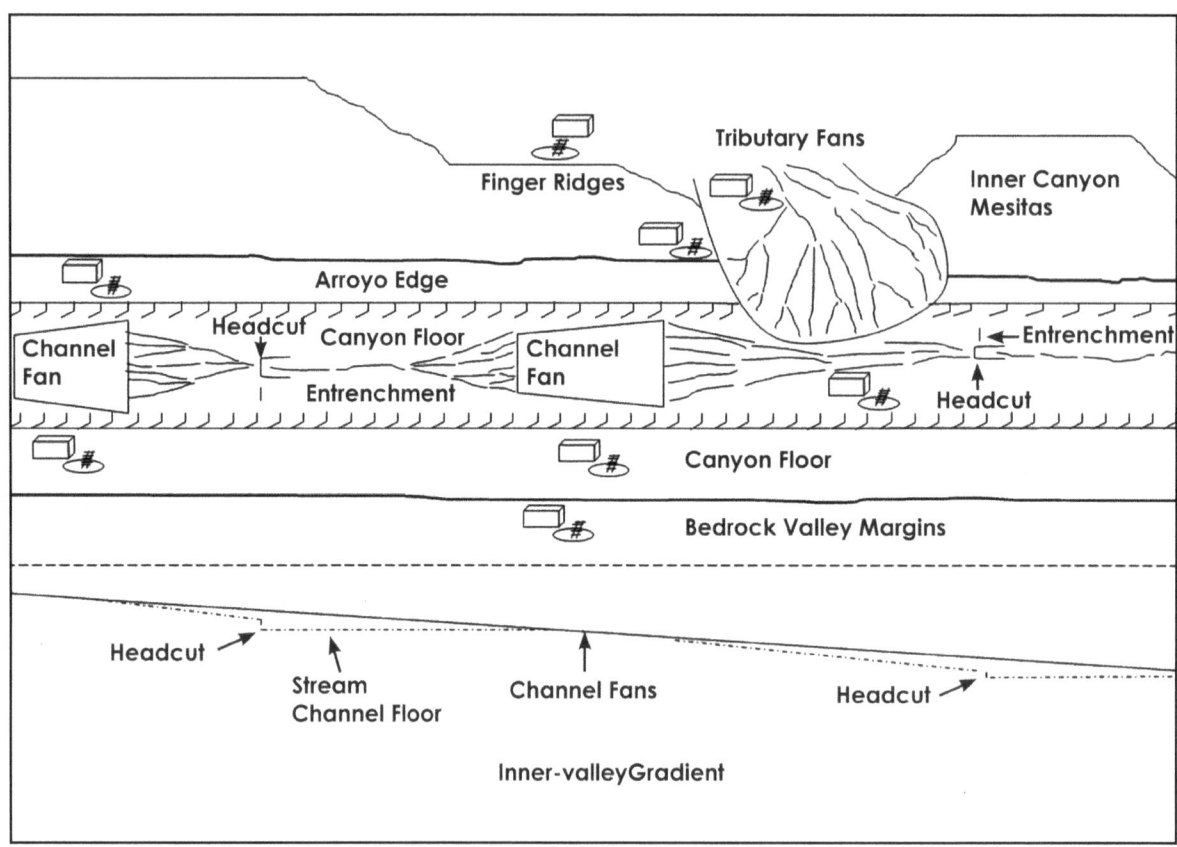

Figure 4.2. A posible Pueblo II–Pueblo III cultural landscape. Important agricultural lands would have existed at aggradation nodes on the active floodplain within the old arroyo and on tributary alluvial fans. Settlements were often located on the Pearson terrace arroyo cutbanks adjacent to the active floodplain, on tributary alluvial fans and on nearby bedrock landforms and finger ridges (the diagram depicting the cutting and filling process along the stream channel is adapted from Huckleberry and Billman, 1998, Figure 8).

Chapter 5
Conclusions and Implications

Findings on Geomorphic History and Puebloan Agriculture

This study adds to our growing body of understanding regarding the relationship of floodplain geomorphic history and the Puebloan agricultural adaption to it in the northern San Juan drainage. In the context of the arroyo-climate debate, we observed only one significant erosion event during the period of Puebloan occupation that would have had a significant impact on agricultural potential. This incision event likely occurred late in PI time but not later than middle PII time and is roughly contemporaneous with what Force and Howell observed in McElmo Canyon (1996) and what Fuller (1988) suggested for the Ute Mountain pediment, an interpretation generally supported by Huckleberry and Bilman (1998).

While we do not suggest a regional chronological sequence nor posit a cause for erosion, we note that the event did result in a lowered water table and reduced area of land where water could be gravity fed to fields, a core tenet of the arroyo-climate model (Bryan 1929, 1941). Several locales in our study area allow us to estimate the amount of irrigable land reduction, with from one-third to one-half of the valley bottom left perched above the lowered water table and likely unsuitable for growing crops.

This event may have produced negative conditions for agriculture while incision was active, but the situation would have been short-lived and very limited, as the subsequent complex response shaped the new floodplain and created conditions where water flow could be delivered to arable land in some stretches of canyon.

However, our direct observations of the new floodplain evolving within the confines of the old arroyo contradict one situation hypothesized by Larry Benson and Michael Berry in their 2009 study: "Farming the post-incision valley floor is risky given that the reduced valley width increase water velocity and level such that "normal" summer convective events will often result in flooded crops" (Benson and Berry, 2009:95). While the threat of flash floods would certainly present problems on a periodic basis (Montezuma Canyon has a history of catastrophic flood events, usually occurring in July and August when crops would have been maturing), our observations indicate that the Puebloan farmers chose to live with that risk, and that they did so successfully for generations.

Buried agricultural features in main channel deposits indicate that infrastructure was regularly lost to floodplain aggradation, but the investment was apparently worth the cost of living on the productive floodland environment. And the catastrophic events such as those that destroyed habitations at the tributary canyon fans at Bradford Canyon, Tank Canyon, and Cave Canyon would certainly have destroyed nearby fields. But again, the floods did not cause the Puebloans to abandon the fan surfaces for settlement and farming, as these places continued to be utilized by subsequent generations.

This moveable aggradation locus, in combination with the fixed tributary canyon alluvial fans, appears to have created an enduring agricultural niche in Montezuma Canyon, as settlement surveys and detailed histories for several sites indicate settlements that endured in place for most, if not all, of Puebloan time.

Lowland Puebloan Historical Ecology and Population Dynamics

This study also provides information useful for informing regional population modeling efforts, most notably the work of the VEP group working in nearby Montezuma Valley, Colorado (Bocinsky and Kohler 2014; Kohler et al. 2008; Kohler et al. 2010; Ortman et al. 2007; Varien et al. 2007), and the model developed by Berry (1982) and more recently elaborated by Benson and Berry (2009).

Focusing on a 1817 km² study area in SW Colorado, the VEP group distilled data from more than 8,948 Puebloan sites into a refined database for 3,176 of the sites that allowed for construction of a fine-grained chronology of 14 periods ranging from 20 to 125 years in length (the majority are 40 years long, or two human generations). Using a hierarchal set of site types to generate household and population estimates, and combining climatic data and soil maps to generate rainfall-fed maize-agriculture productivity estimates, the VEP group constructed a 700 year occupational history of their upland study area. This model identified two cycles of population growth followed by steep declines: one in the 800s, which falls within the PI period, and a second, more dramatic peak during the PIII period, which saw the growth of the largest Puebloan villages in the Mesa Verde area but also followed by the sudden and dramatic abandonment of the entire region. The VEP study concludes that these population cycles were driven by changing rainfall and temperature patterns, which, at times, provided conditions favorable for rain-fed maize agriculture, but interrupted by episodes of severe drought and on occasion cold snaps, with drought being the main driver of out-migration. However, the VEP researchers recognize that their modeling exercise is specific to their upland study area and that to understand the full historical ecology of the Puebloan peoples, there is a need to work at a larger scale (Varien et al. 2007: 294).

We believe that Montezuma Canyon can provide critical information for understanding population dynamics on that larger scale. Although the VEP study provides a thorough and compelling analysis of maize agricultural potential in an upland setting, the study's western boundary falls for the most part along the escarpment of the Great Sage Plain Plateau, where rain-fall dependent agriculture was still feasible, but their exercise does not take into consideration the Puebloan populations that adapted to life below 1600 m (5200 ft) elevation, the accepted moisture threshold below which rainfall-dependent maize agriculture could be practiced in the northern San Juan (Shaw 1988).

We believe the VEP model would benefit by including settlement and geomorphological data from the low elevation floodwater environment of Montezuma Canyon, and we turn to a discussion of the VEP study's period 5 (A.D. 880 to 920) and period 6 (A.D. 920 to

980) to illustrate our point.

The early period of population growth and village aggregation documented in the VEP study area began during their modeling period 3 (A.D. 800 to 840) and reached its peak in period 4 (A.D. 840 to 880). Montezuma Canyon was also occupied at this time, with PI settlements located at the mouths of Dodge, Coalbed, Tank, Monument, and Nancy Patterson canyons, among many others. This period of population growth and aggregation into large villages came to an end in the VEP study area during their modeling period 5 (A.D. 880 to 920), a time marked by a prolonged cold drought (Petersen 1988; Kohler et al. 2008; Ortman et al. 2007).

However, rather than witness dramatic out migration (or population decline) as occurred in the VEP study area, Montezuma Canyon appears to have experienced a stable, if not increasing, population, as evidenced by the late PI/early PII components such as found at Nancy Patterson Village (Janetski and Hurst 1984; Janetski and Thompson 2012), Cave Canyon Village (Christensen 1980; Harmon 1977; Neilsen 1978), and Three Kiva Pueblo (D. Miller 1974), these being three of the four sites excavated by the BYU field schools. With more than sufficient frost-free days to accommodate maize cultivation, and what this study suggests were conditions suitable for runoff-based ak-chin agriculture on main valley channel fans and tributary canyon fans, Montezuma Canyon may have provided a suitable agricultural niche for Puebloan populations to remain in place during the early 900s, if not absorb emigrants from surrounding uplands. Although the VEP team postulates that a north-to-south migration to the Rio Grande valley, the Tucson Basin, or closer to home, "into northwestern New Mexico and the San Juan geologic Basin" (Varien et al. 2007: 289), a compelling case could be made for a much shorter horizontal migration via a "vertical" migration into the nearby floodplain of Montezuma Canyon.

The population model proposed by Michael Berry in his 1982 study bolsters this argument. In his study, Berry argued that the phase transitions in the Pecos classification of Southwestern prehistory were real events, not just classificatory conveniences, and that the cultural transitions resulted from Plateau-wide droughts which impacted maize agricultural potential and caused population movements and aggregations into several discrete agricultural refugia. By combining an analysis Puebloan culture history, chronometric data, and climate history, Berry was able to make a compelling case (albeit controversial, see Dean 1985; Irwin-Williams 1985) for drought driven population movements and refugia-based cultural interactions as the genesis for the cultural changes marking phase transitions in the Pecos system.

Subsequently, Larry Benson and Michael Berry (2009) have employed an enhanced set of climatic and chronological data to further the argument that drought and its impact on maize agricultural potential was the driving force in shaping Puebloan culture history. In their study, they identify six mega droughts (dry period lasting more than 20 years with shorter consecutive-year spans within those decades with yet more severe dry conditions). For our discussion, we will look at their mega drought "D1," which lasted from A.D. 863 to 884 (Benson and Berry 2009: 106, Figure 8.B), and which provides an interesting comparison with the findings of the VEP study for the same period.

Although Berry (1982:114) recognized the transition from PI to PII as "the greatest discontinuity in the post-Basketmaker Anasazi record" (aside, of course, from the A.D. 1300 abandonment), in his original analysis of sites typical of the PI and PII village patterns, Berry found records for two sites in the Canyon de Chelly area that represent an anomaly in the sequence, as both appear to have been occupied across the phase transition boundary. Oddly, the two neighboring Arizona sites lie in very different ecological settings. One site, NA8013 of the Cross Canyon Group (Olson 1971, as cited in Berry 1982), is an upland site on the Defiance Plateau at about 2135 m (7000 ft), a high elevation location that would have received greater moisture but was near the upper limit of the cold tolerance threshold for maize agriculture. The other site, Sliding Ruin, is a cliff dwelling located below 1700 m elevation (5600 ft) in Canyon de Chelly, adjacent to Chinle Wash. Chinle Wash drains a large expanse of the Defiance Plateau, with elevations consistently above 2400 to 2700 m (8,000 to 9,000 ft), but Sliding Ruin was located in a lowland setting to take advantage of abundant frost free days in a niche where there was apparently adequate runoff to water fields.

Berry postulates that these two communities found distinct agricultural niches that allowed them to persist in place while others failed. He also recognizes that these two sites are likely not the only ones that survived the drought and suggests that more survey in the Defiance Plateau would likely find more. We would argue that Montezuma Canyon and its adjoining uplands is also a logical place to look, and refer to two local sites that fit the Defiance Plateau pattern. Nancy Patterson Village lies at the confluence of Cross Canyon at an elevation of 1460 m (4800 ft), and would have relied entirely on runoff flowing into the confluence of Cross and Montezuma canyons to water its agricultural lands. On the same Cross Canyon drainage, some 35 km to the east, is Greenlee Village (Dove et al. 2006; Dove 2012), situated on the edge of the Great Sage Plain Plateau near the head of Bug Canyon above 2015 m elevation (6600 ft), a zone where there was sufficient rainfall to sustain agriculture but, near the zone, limited by the number of growth-degree days for maize maturation. Although they are more than 550 m (1800 ft) apart in elevation, both villages have components that date to the PI-PII transition, indicting they survived the mega drought of the late 800s while occupying very different ecological niches.

Other sites located in similar settings, such as the aforementioned villages in Montezuma Canyon and the cluster of upland sites documented in the Dove Creek area by Grant Coffey (2004), indicate a bi-modal settlement strategy that allowed Puebloan communities to persist in place through difficult environmental times. While the VEP climatic and soils data can be used to understand how these communities were able to persist in the uplands, we still need to understand how conditions on the Montezuma Canyon floodplain influenced the lowland adaptation. Montezuma Canyon provides the sediments and sites for such a study.

IMPLICATIONS FOR MANAGEMENT AND FUTURE RESEARCH

As this and other studies have demonstrated, Montezuma Canyon contains a wealth of Puebloan archaeological sites set in the context of well-preserved, well-exposed, accessible, and mapable depositional units. This unique juxtaposition is important for future research and also provides an important teaching tool for students as well as land managers, but more effort is needed in certain areas.

As mentioned at the outset, this study is a reconnaissance level effort to understand the basic elements of the depositional history of the canyon. As such, we were limited in the number of places we could visit and the amount of time we could spend at each. However, all of these 17 study sites are accessible for further study, and more effort expended to examine cut banks beyond those we found would further refine our understanding. Greater effort looking at the depositional history and Puebloan land use near known archaeological sites should be particularly productive in this effort. The study could also be expanded to look at depositional history up some of the side tributaries, particularly Cross Canyon, as well as continue the study into the lower canyon onto Navajo Nation lands all the way to the San Juan River. It should also be expanded north of Pearson Canyon to the rim rocks at the head of the main stems of Montezuma Canyon.

As our study demonstrated, we were able to map unit surface distributions in certain areas, but time did not allow us to complete the work. It should be possible to generate a map of Holocene depositional units for most of the canyon. This would be particularly valuable for land managers, as depositional history is the primary determinant in whether to anticipate buried cultural materials. For example, where the Pearson Unit comprises the canyon floor, cultural materials would be expected to be found on the surface and buried materials rare. However, where the Horsehead Unit or the Nancy Patterson Unit comprise the surface, extra care should be taken to explore for buried materials as these two units could encapsulate or conceal them. This understanding could anticipate inadvertent discoveries such as the buried BMII site discovered during excavation for a pipeline near the mouth of Cave Canyon (Hovezak and Harden 1987), where we would surmise colluvium concealed the Pearson Unit surface where the site was located on the canyon margin.

More research could also reveal much about the canyon's past environments and depositional history. As charcoal is common, if not abundant, throughout Pearson and Horsehead unit deposits, an effort should be made to identify and collect species-specific charcoal samples (good contexts, short-lived species, etc.) that can be used to reliably date depositional units. The sediments are also sure to contain pollens that will indicate past plant communities, both natural and anthropogenic.

The Pearson unit in particular has the potential to contain pollens that reflect plant communities that were present during the mid-to-late Holocene. The search for corn pollen in this unit might prove particularly interesting, as its mere presence would indicate the arrival of agriculturalists in the larger drainage basin even though their settlement sites might not yet appear in the archaeological record (see Wahl et al. 2006, 2007a, 2007b for an example from the Maya lowlands).

Work to systematize the archaeological database is also needed. Information on the many thousands of archaeological sites found in the Montezuma Canyon drainage documented over the past 140 years represents a highly variable but difficult to interpret record. Sites have been recorded using different standards and methodologies. For example, Kenneth Wintch's 1990 effort to construct population aggregation and movements using the last and best iteration of site recording forms — the IMACS — proved unsuccessful, not for lack of apparent patterns in the archaeological record, rather for lack of consistency in site documentation (Wintch 1990:86). The VEP researchers encountered the same problem in

Colorado but through a systematic strategy were able to wrangle disparate data on several thousand sites into a manageable format (Ortman et al. 2007). Application of the VEP strategy onto the Montezuma Canyon site records should prove equally worthwhile.

Gaining an understanding of the relationship of climate to floodplain evolution, and the relationship of Puebloan farmers to both, has proven a vexing problem for generations of researchers. As this study has demonstrated, Montezuma Canyon provides an ideal setting for unraveling this complex relationship in the northern San Juan Basin. It is only when we understand how Puebloan peoples adapted to life in this precarious lowland environment during times of drought that emptied the uplands of people will we understand the full spectrum of their overall achievement of surviving for centuries in the northern San Juan drainage.

References Cited

Antevs, E.
    1952   Arroyo-Cutting and Filling. *The Journal of Geology* 60: 375-385.

Armstrong, R.W.
    1969   K-Ar dating of laccoithic centers of the Colorado Plateau and vicinity. In *Geological Society of America Bulletin*, Vol. 80, pp. 2081-2086.

Baer, Sarah
    2003   *Temporal Change in the Community Centers of the Anasazi of Montezuma Canyon, Utah – AD 1050-1200*. Unpublished Master's thesis, Department of Anthropology, Brigham Young University, Provo.

Benson, Larry V., and Michael S. Berry
    2009   Climate Change and Cultural Response in the Prehistoric American Southwest. *Kiva* 75(1).

Berry, Michael S.
    1982   *Time, Space, and Transition in Anasazi Prehistory*. University of Utah Press, Salt Lake City.

Betancourt, J., and T. Van Devender
    1981   Holocene Vegetation in Chaco Canyon, New Mexico. *Science* 214:656-658.

Billat, Lorna
    1985   Homesteading and trading in Montezuma Canyon: A Frontier Necessity. Unpublished Master's thesis, Department of Anthropology, Brigham Young University, Provo.

Bocinsky, R. K., and Timothy A. Kohler
    2014   A 2,000 year reconstruction of the rain-fed maize agricultural niche in the US Southwest. Nat. Commun. 5:5618 doi: 10.1038/ncomm6618.

Bond, Mark
    1984   *Archaeological Survey of the San Juan County Road's Proposed Country Road 146 Realignment at the Confluence of Tank and Montezuma Canyons, San Juan County, Southeastern Utah*. Abajo Archaeology, Bluff, Utah.

Bryan, K.
    1929   Floodwater Farming. *Geographical Review* 19:444-456.

    1941   Pre-Columbian Agriculture in the Southwest, as Conditioned by Periods of Alluviation. *Annals of the Association of American Geographers* 31:219-242.

Christensen, Diana
    1980   Excavation at Cave Canyon Village, 1977, Montezuma Canyon, Utah. Unpublished Master's thesis, Department of Anthropology, Brigham Young University, Provo, Utah.

Christenson, G.E.
    1985   Quaternary Geology of the Montezuma Creek-Lower Recapture Creek Area, San Juan County, Utah. In *Contributions to Quaternary Geology of the Colorado Plateau*, pp. 3-31. Utah Geological and Mineral Survey Special Studies 64.

Coffey, G.D.
    2004   Regional Migration and Local Adaptation: A Study of Late Pueblo I and Early Pueblo II Sites in the East Dove Creek Area. Unpublished Master's thesis, Northern Arizona University, Flagstaff.

Cooke, R.U., and R. W. Reeves
    1976   *Arroyos and Environmental Change*. Clarendon Press, Oxford, England.

Cordell, L.S.
    1975   Predicting Site Abandonment at Wetherill Mesa. *The Kiva* 40:189-202.

Dean, Jeffrey S.
    1985   Review of Time, Space, and Transition in Anasazi Prehistory, by M. S. Berry. *American Antiquity* 50:704–705.

    1988   A Model of Anasazi Behavioral Adaptation. In *The Anasazi in a Changing Environment*, edited by George J. Gummerman, pp. 25–44. Cambridge University Press, Cambridge.

Dean, Jeffrey S., Robert C. Euler, George J. Gumerman, Fred Plog, Richard H. Hevly, and Thor N. V. Karlstrom
    1985   Human Behavior, Demography, and Paleoenvironment on the Colorado Plateaus. *American Antiquity* 50:537–554.

Dean, Jeffrey S., and Carla R. Van West
    2002   Environment-Behavior Relationships in Southwestern Colorado. In *Seeking the Center Place: Archaeology and Ancient Communities in the Mesa Verde Region*, edited by Mark D. Varien and Richard H. Wilshusen. University of Utah Press, Salt Lake City.

de Haan, Petrus. A.
    1972   An Archaeolgocial Survey of Lower Montezuma Canyon, Southeast Utah. Unpublished Master's thesis, Department of Anthropology, Brigham Young University, Provo, Utah.

Dove, David M.
    2012   Ritualized Animal Use in an Early Kiva: Champaign Spring (Greenlee) Ruin 4DL2333. Unpublished manuscript.

Dove, Donald E., Steven Dinaso, David Dove, Kimberly M. Gerhardt, Vincent P. Gutowski, Harvey Henson, Patricia F. Lacey, Diane McBride, Robert McBride, Jonathan Till and Larry Tradlener
    2006   Topographical Mapping, Geophysical Studies and Archaeological Testing of an Early Pueblo II Village near Dove Creek, Colorado. Unpublished manuscript.

Eiselt, B. Sunday, J. Andrew Darling, Samuel Duew, Mark Willis, Chester Walker, William Hudspeth, and Leslie Reeder-Meyers
    2017   A Bird's Eye View of Proto-Tewa Subsistence Agriculture: Making the Case for Floodplain Farming in the Ohkay Owingeh Homeland, New Mexico. *American Antiquity* 82:397-413.

Euler, R.C.
    1988   Demography and Cultural Dynamics on the ColoradoPlateau. In *The Anasazi in a Changing Environment*, edited by G. J. Gumerman, p. 192-239. Cambridge University Press, Cambridge.

Euler, Robert C., George Gumerman, Thor N. V. Karlstrom, Jeffrey Dean, and Richard Hevly
    1979   The Colorado Plateaus: Cultural Dynamics and Paleoenvironment. *Science* 205:1089–1101.

Fish, S.K., P.R. Fish, J.H. Madsen
　　1994　Toward an Explanation of Southwestern "Abandonments". In *Themes in Southwestern Prehistory*. Edited by G.T. Gumerman, pp. 135-163. School of American Research, Santa Fe.

Force, E. R., and W. K. Howell
　　1996　*Holocene Depositional History and Anasazi Occupation in McElmo Canyon, Southwestern Colorado*. Arizona State Museum Archaeological Series, No. 188. University of Arizona Press, Tucson.

Fuller, S.L.
　　1988　*Cultural Resources Inventory for the Dolores Project: The Ute Irrigated Lands Survey*. Four Corners Archaeological Project Report No. 13. Complete Archaeological Services Associates, Cortez, Colorado.

Hack, J.T.
　　1942　The Changing Physical Environment of the Hopi Indians of Arizona. *Papers of the Peabody Museum* 35(1). Harvard University, Cambridge.

Harmon, Craig B.
　　1977　Cave Canyon Village (42Sa2096): Excavations in the Early Pueblo Components, Montezuma Canyon, Sand Juan County, Southeastern Utah. Unpublished Master's thesis, Department of Anthropology, Brigham Young University, Provo, Utah.

　　1979　*Cave Canyon Village: The Early Pueblo Components*. Publications in Archaeology, New Series No. 5. Department of Anthropology, Brigham Young University, Provo, Utah.

Haynes, C.V., Jr.
　　1968　Geochronology of Late-Quaternary alluvium. In *Means of Correlation of Quaternary Successions; Proceedings 7th INQUA Congress*, edited by R. B. Morrison & H. E. Wright, pp. 591-631. University of Utah Press, Salt Lake City.

Hovezak, Timothy and Patrick L. Harden
　　1987　Excavations at 42Sa17440, A Basketmaker II Site in Montezuma Canyon, San Juan County, Utah. *LAC Report 8598b*, La Plata Archaeological Consultants, Inc. Dolores, Colorado

Huckleberry, Gary A., and Brian R. Billman
　　1998　Floodwater Farming, Discontinuous Ephemeral Streams, and Puebloan Abandonment in Southwestern Colorado. *American Antiquity* 63:595–616.

Huff, L.C., and F.G. Lesure
　　1965　Geology and Uranium Deposits of Montezuma Canyon Area, San Juan County, Utah. *U.S Geological Survey Bulletin* 1190.

Hurst, Winston, and Joel C. Janetski
　　1985　*The Nancy Patterson Village Archaeological Project: Field Year 1984 – Preliminary Report No. 2*. Technical Series No. 85-32. Museum of People and Cultures, Brigham Young University, Provo, Utah.

Irwin-Williams, Cynthia
　　1985　Review of "Time, Space and Transition in Anasazi Prehistory", by Michael S. Berry, University of Utah Press. *The Kiva* 51(1): 44-48.

Jackson, William H.
　　1878　Report on the Ancient Ruins Examined in 1875 and 1877. *US Geological and Geographical Survey of the Territories for 1876*, 10th Annual Report, pp. 411-450. Washington, D.C.

Janetski, Joel C., and Winston Hurst
    1984   *The Nancy Patterson Village Archaeological Research Project: Field Year 1983 – Preliminary Report.* Technical Series No. 84-7. Museum of People and Cultures, Brigham Young University, Provo, Utah.

Janetski, Joel C., and Charmaine Thompson
    2012   Puebloan Subsistence in Montezuma Canyon, Southeast Utah. In *An Archaeological Legacy: Essays in Honor of Rat T. Matheny.* Occasional Paper No. 18, Museum of Peoples and Cultures, Brigham Young University, Provo.

Karlstrom, T.N.V.
    1988   Alluvial Chronology and Hydrologic Change of Black Mesa and Nearby Regions. In *The Anasazi in a Changing Environment*, edited by G. J. Gumerman, pp. 45-91. Cambridge University Press, New York.

Karlstrom, E.T. and T.N.V. Karlstrom
    1987   Late Quaternary Alluvial History of the American West: Toward a Process Paradigm. *Geology* 15:88-89.

Kohler, Timothy A.
    1992   Prehistoric Human Impact on the Environment in the Upland North American Southwest. *Population and Environment: A Journal of Interdisciplinary Studies* 13:255-268.

Kohler, Timothy A., C. David Johnson, Mark D. Varien, Scott G. Ortman, Robert G. Reynolds, Ziad Kobti, Jason Cowan, Kenneth Kolm, Schaun Smith, and Lorene Yap
    2007   Settlement Ecodynamics in the Prehispanic Central Mesa Verde Region. In *Modeling Socionatural Systems,* edited by Timothy A. Kohler, and Sander van der Leeuw, pp. 61–104. SAR Press, Santa Fe.

Kohler, Timothy A., Matt Pier Glaude, Jean-Pierre Bocquet-Appel and Brian M.Kemp
    2008   The Neolithic Demographic Transition in the U.S. Southwest. *American Antiquity* 73(4):645-669.

Kohler, Timothy A., and Mark D. Varien
    2012   *Emergence and Collapse of Early Villages: Models of Central Mesa Verde Archaeology.* University of California Press, Berkeley.

Kohler, Timothy A., Mark D. Varien and Aaron Wright
    2010   Leaving Mesa Verde: Peril and Change in the Thirteenth Century Southwest. University of Arizona Press, Tucson.

Lipe, William D.
    1992   The Depopulation of the Northern San Juan: Conditions in the Turbulent 1200s. *Journal of Anthropological Archaeology* 14:163-169.

Loosle, Byron
    1988   *Montezuma Canyon Projectile Points as Temporal Markers for the Mesa Verde Anasazi.* Unpublished Master's Thesis, Department of Anthropology, Brigham Young University, Provo, Utah.

Matheny, Ray T.
    1962   An Archaeological Survey of Upper Montezuma Canyon, San Juan County, Utah. Unpublished Master's thesis, Department of Anthropology, Brigham Young University, Provo, Utah.

Miller, Blaine
    1976   A Study of a Prudden Unit Site (42Sa971-N) in Montezuma Canyon, San Juan County, Utah. Unpublished Master's thesis, Department of Anthropology, Brigham Young University, Provo, Utah.

Miller, Donald E.
    1974   A Synthesis of Excavations at Site 4Sa863, Three Kiva Pueblo, Montezuma Canyon, San Juan County, Utah. Unpublished Master's thesis, Department of Anthropology, Brigham Young University, Provo, Utah.

Montoya, Donald G.
    2008   A Hidden Village (42Sa2112): A Basketmaker III Community in Montezuma Canyon, Utah. Unpublished Master's thesis, Department of Anthropology, Brigham Young University, Provo.

Nielsen, Glenna
    1978   Excavations in the Basketmaker Component of Cave Canyon village: (Site 42Sa2096) 1976 Field Season, Montezuma Canyon, San Juan County, Utah. Unpublished Master's thesis, Department of Anthropology, Brigham Young University, Provo, Utah.

Ortman, Scott G., Mark D. Varien, and T. Lee Gripp
    2007   Empirical Bayesian Methods for Archaeological Survey Data: An Application from the Mesa Verde Region. *American Antiquity* 72:241-272.

Patterson, Gregory R.
    1975   A Preliminary Study of an Anasazi Settlement (42Sa971) Prior to A.D. 900 in Montezuma Canyon, San Juan County, Southeastern Utah. Unpublished Master's thesis, Department of Anthropology, Brigham Young University, Provo, Utah.

Patton, P.C. and S.A. Schumm
    1981 Ephemeral Stream Processes; Implications for the Studies of Quaternary Valley Fills. *Quaternary Research* 15:24-43.

Petersen, Kenneth Lee
    1988   *Climate and the Dolores River Anasazi: A Paleoenvironmental Reconstruction from a 10,000-Year Pollen Record, La Plata Mountains, Southwestern Colorado.* Anthropological Papers No. 113. University of Utah, Salt Lake City.

Schlanger, Sarah H.
    1988   Patterns of Population Movement and Long-Term Population Growth in Southwestern Colorado. *American Antiquity* 53:773–793.

Schwindt, Dylan M., R. Kyle Bocinsky, Scott G. Ortman, Donna M. Glowacki, Mark D. Varien, and Timoth A. Kohler
    2016   The Social Consequences of Climate Change in the Central Mesa Verde Region. *American Antiquity* 81(1):74-96.

Schumm, S.A.
    1977   *The Fluvial System.* John Wiley and Sons, New York.

Schumm, S.A., and R.S. Parker
    1973   Implications of Complex Response of Drainage Systems for Quaternary Alluvial Stratigraphy. *Nature* 243:99-100.

Shaw, Robert H.
   1988   Climate Requirement. In *Corn and Corn Improvement*, edited by G.F. Sprague and J.W. Dudley, pp. 609-638. American Society of Agronomy, Madison.

Van West, Carla, R.
   1994   *Modeling Prehistoric Agricultural Productivity in Southwestern Colorado: A GIS Approach*. Reports of Investigations No. 67. Department of Anthropology, Washington State University, Pullman, and Crow Canyon Archaeological Center, Cortez, Colorado.

Van West, Carla and William D. Lipe
   1995   Modeling Prehistoric Agriculture and Climate in Southwestern Colorado. In *The Sand Canyon Archaeological Project: A Progress Report*, edited by William D. Line, pp. 105-119. Occasional Paler No. 2, Crow Canyon Archaeological Center, Cortez.

Varien, Mark D., and Richard H. Wilshusen (eds.)
   2002   *Seeking the Center Place: Archaeology and Ancient Communities in the Mesa Verde Region*. University of Utah Press. Salt Lake City.

Varien, Mark D., Scott G. Ortman, Timothy A. Kohler, Donna M. Glowacki, and C. David Johnson
   2007   Historical Ecology in the Mesa Verde Region: Results From the Village Ecodynamics Project. *American Antiquity* 72(2), 2007, pp. 273-299.

Wahl, David, Roger Byrne, Thomas Shreiner, and Richard Hansen
   2006   Holocene vegetation change in the northern Peten and its implications for Maya Prehistory. *Quaternary Research* 65:380-389.

   2007a   Paleolimnological Evidence of late-Holocene Settlement and Abandonment in the Mirador Basin, Peten, Guatemala. *The Holocene* 17(6):813-820.

Wahl, David, Thomas Schreiner, Roger Byrne, and Richard Hansen
   2007b   A Paleoecological Record from a Maya Reservoir in the North Peten. *Latin American Antiquity* 18:212-222.

Wilde, James D. and Charmaine Thompson
   1988   Changes in Anasazi Perceptions of Household and Village Space at Nancy Patterson Village. *Utah Archaeology* 1(1):29-45.

Wintch, Kenneth L.
   1990   Patterns of Pueblo Aggregation and Dispersal: A Test Case from the Montezuma Creek Drainage in San Juan County, Utah. Unpublished Master's thesis, Department of Anthropology, Brigham Young University, Provo, Utah.

www.ingramcontent.com/pod-product-compliance
Lightning Source LLC
Chambersburg PA
CBHW061141230426
43663CB00027B/2991